Digital Culture Shock

Digital Culture Shock

WHO CREATES TECHNOLOGY AND
WHY THIS MATTERS

KATHARINA REINECKE

PRINCETON UNIVERSITY PRESS
PRINCETON AND OXFORD

Published by Princeton University Press
41 William Street, Princeton, New Jersey 08540
99 Banbury Road, Oxford OX2 6JX

press.princeton.edu

GPSR Authorized Representative: Easy Access System Europe - Mustamäe tee 50,
10621 Tallinn, Estonia, gpsr.requests@easproject.com

All Rights Reserved

ISBN 978-0-691-25581-1
ISBN (e-book) 978-0-691-27029-6

Library of Congress Control Number 2025935251

British Library Cataloging-in-Publication Data is available

Editorial: Hallie Stebbins and Chloe Coy
Production Editorial: Jill Harris
Jacket Design: Hunter Finch
Production: Erin Suydam
Publicity: Matthew Taylor and Kate Farquhar-Thomson
Copyeditor: Bhisham Bherwani

This book has been composed in Arno and Sans

Printed in the United States of America

10 9 8 7 6 5 4 3 2 1

For Malte and Lily
May you witness all the world's cultural diversity

culture shock, n. [ˈkʌltʃər ʃɑːk]: a sense of confusion and uncertainty sometimes with feelings of anxiety that may affect people exposed to an alien culture or environment without adequate preparation.

—*Merriam-Webster Dictionary*

CONTENTS

Acknowledgments   xi

1   Introduction   1

2   Kaleidoscope of Cultures   24

3   Perceiving Information in Different Cultures   44

4   Technology Use around the World   61

5   Use of Online Communities across Cultures   88

6   Tell Me Where You Live and I'll Tell You
        What You Like   102

7   Communicating with and
        through Technology   120

8   Culture Shocked by Technology   145

9   Cultural Imperialism and Marginalization
        through Technology   161

10 Building Culturally Just Infrastructure   178

11 Epilogue   198

Bibliography   205
Index   223

ACKNOWLEDGMENTS

I had plenty of realizations while writing this book, but one is that I'm incredibly grateful to many more people than I will ever be able to fit into a short acknowledgment. I'll give it a try anyway:

All the really smart colleagues who, through discussions, reviews, and feedback helped me improve the book and gave me the courage to start and finish it: Ishtiaque Ahmed, Rodolfo Barragan, Eshwar Chandrasekharan, Dipto Das, Batya Friedman, Pamela Hinds, Sandy Kaplan, Neha Kumar, Sebastian Linxen, Jennifer Mankoff, Andrew Meltzoff, Tanu Mitra, Michael Muller, Jacki O'Neill, Daniel Russell, Bryan Semaan, Christian Sturm, Kentaro Toyama, Aditya Vashistha, Naomi Yamashita, the anonymous reviewers, and many more. Thank you so much.

My amazing team of book debuggers: Jeffrey Basoah, Silvia Lindtner, Joyojeet Pal, Daniela Roesner, and Thomas Talhelm. I cannot thank you enough for making my book workshop so productive and enjoyable and for letting me see more than just "the peas on my plate." Silvia, thank you for taking me on a book-publishing crash course over Zoom, text messages, and a was-it-really-eight-courses dinner in Hamburg.

All the people supporting me in my interest in culture: My parents, who gave me every opportunity to visit different countries; my PhD advisor, Avi Bernstein, who offered me complete freedom to make my own research path; Shinobu Kitayama for the coffees and many conversations about cultural psychology; and my postdoc advisor, Krzysztof Gajos, for the many walks and talks, and for the amazing friendship.

My students, who never cease to teach me and who have contributed to much of the research presented in this book: Tal August, Jeffrey Basoah, Elizabeth Castillo, Alice Gao, Jeongyeob Hong, Noah Huck,

Katherine Juarez, Qisheng Li, Rohit Maheshwari, Katelyn Mei, Nino Migineishvili, Drishya Nair, Rock Yuren Pang, Sebastin Santy, Kyle Thayer, Yadi Wang, Spencer Wiliams, Judith Yaaqoubi, Julie Yu, Jiawen Stefanie Zhu, and the rest of the Wildlab. I have learned so much from you.

Nigini Oliveira, for your endless enthusiasm about LabintheWild, for your dedication to making science less WEIRD, for cheering me up with stories about compost toilets, tiny fireplaces, and Stuart, the cat, and for your friendship.

My incredibly supportive Princeton University Press team: Hallie Stebbins, who embraced the idea behind this book from the start to the end and cheered me on throughout the process; Chloe Coy, Susan Clark, and Jill Harris for guiding me through all sorts of steps that I never thought were part of publishing a book; and to Bhisham Bherwani for the excellent copyediting.

The Swiss National Science Foundation, the US National Science Foundation, and the Hasler Foundation, who funded much of the work that I present in this book, and the University of Washington for sponsoring my writing sabbatical.

The LabintheWild participants from all corners of the world: this book would not exist without you.

Sally Engle Merry, who encouraged me to write this book and taught me when it is okay and not okay to give in to the seductions of quantification. You did not live to see me put any words down, but so many parts of the book were influenced by you.

My friends, family, and neighbors, who have listened and supported me with their enthusiasm and questions, and distracted me when I had to get my mind off the book: Ofra Amir, Dimitrios Antos, Jessica Brinton, Maya Cakmak, Cecilia Peralta Ferriz, Catherine Gannon, Raffay Hamid, Cindy Hamra, Marguerite Hutchinson, Steve Komarov, Amy Johnson, Franziska Liese, Sarah Mennicken, Sarah Merry, Josh Merry, Chelsea Miller, Regina Miller, Kris Myllenbeck, Du Nguyen, Lena Schultz, Bonita Sen, Pao Siangliulue, Maggie Stringfellow, Ricca Ranariboana, Gerald Reif, Justin Richland, Lindsey Richland, Carli

Rogosin, Pradeep Suri, Ping Xu, Margaret Young, Laura Zaugg, Liza Zurlinden, and the rest of my "village." I'm so lucky to know all of you.

Finally, thanks to my husband, Sawyer Fuller, for making coffee, for cooking meals, and for telling me that even a paragraph per day is progress. And to my kids, who pointed at any new book that arrived in our house, asking, "Is this your book, Mom?" I guess I had no choice but to see this through.

Digital Culture Shock

1

Introduction

On February 28, 2023, the autonomous driving technology company Waymo announced that its cars had achieved a new milestone: Without anyone in the driver's seat, its robotaxis have traveled more than one million miles on public roads in major US cities—a distance that the average American driver would complete in about 80 years.[271] During that time, an American driver would have likely been involved in three or four accidents, ranging from minor parking lot scrapes to fatal crashes. In comparison to this, the data Waymo released showed that its cars are extremely safe: Over the course of a million miles, they didn't have any accidents that included pedestrians or bicyclists, and they had only two serious collisions in which at least one of the involved cars had to be towed. Of the 20 low-severity collisions, most happened when a Waymo car was stationary or driving very slowly, and most were caused by other drivers or loose objects. In one incident, another car backed out of a parking slot and into an idling robotaxi. In another, a garbage truck tried to squeeze by a Waymo car and didn't quite make it. And in yet another incident, a plastic sign blew across the street and hit one of their robot cars.[286] Even I had to admit that Waymo's safety performance has been arguably better than those of human drivers. In fact, reading about its safety record and looking through the data made me wonder why I've been so skeptical of self-driving cars at all. (There are, of course, good reasons to be cautious, as several serious incidents with self-driving cars have shown.[268])

Part of the reason for my skepticism toward autonomous vehicles is that, as a computer scientist, I have seen how frequently software systems fail and how often human beings are to blame. I have seen how difficult it is to anticipate the myriad of situations that could arise when software is released into the real world. And I have heard about the promises of artificial intelligence, or AI for short, for several decades—only to find out that "intelligence" is a pretty broad term that is hyping the technology at best, and is extremely misleading at worst.

But another reason for my skepticism is that the Waymo story sounds all too familiar to me. Here's a tech company that hopes to address several problems it sees at once, such as that people apparently hate sitting in traffic without being able to watch TV, that those who cannot drive themselves can be severely limited in their mobility, and that humans are horrible drivers who are all too easily distractible, react slowly, and have short attention spans. The company develops a new technology that is supposed to free us of these problems, puts it out into the world, and observes what happens. Its conclusion? People readily hail their robotaxis, only very few complain, and the overall safety performance is great, for the most part.

Except that it isn't that easy. All we know right now is that only a tiny portion of Americans are ready to hand over control to a robot, and that these are a minority even in San Francisco and a few other urban centers where the robotaxis have been tested. Others may not share the same trust in artificial intelligence; as a matter of fact, a global survey of 17,193 people published in 2023 by researchers at the University of Queensland, Australia,[99] found that only 16% of the participants from Finland and 23% of those from Japan reported trusting AI. Across the 17 countries that the researchers surveyed, three out of five participants reported being ambivalent about or unwilling to trust AI. Additionally, many people may simply not think the lack of self-driving cars is a problem, because it only is if you live in a car-dominated place. And even more important than who may or may not be adopting this technology is that we don't know whether it will actually work in other places.

So here's an intriguing question that I think is worth entertaining. What would happen if US-designed robotaxis were deployed in countries outside of the US? Having grown up in Germany, I believe many people there would throw their hands up in frustration to see an American robocar defensively inch forward at every intersection just because the US decided to be one of very few countries in the world to employ four-way stops throughout its cities. The US has rules as to who can go first in these intersections, but first and foremost the rule seems to be to go slow and give the right-of-way to the more offensive drivers. And this issue of communicating between cars and drivers is huge: Just imagine how an autonomous car would signal to the driver of another car that they should go first through the intersection. Would it be best if it flashed its lights? If it backed up a bit? Or if it extended and waved a robotic arm at the other driver? How to insert more human-like behavior into robots is a difficult question, but one that researchers have feverishly worked on over the past decades. And much progress has been made: For example, a team from the University of California, Berkeley, found a way to make robot cars more courteous.[261] They modeled human behavior based on a dataset of people's driving behavior on a highway near Emeryville, California, which had been recorded by surrounding video cameras. How do people behave when changing to another lane? Do they speed up, slow down, or simply pull over? Analyzing this dataset, the team of researchers found that people are generally very courteous: Whereas a selfish driver would overtake another car by simply changing lanes, no matter whether this would cause another driver to severely hit the brake, most people slowed down to let a car pass first before shifting into the new lane. When the researchers simulated such courteous behavior in a robot car trying to merge in front of another, they found it left as much space for the following car as human drivers would. Ultimately this would mean that robot cars would drive more safely with this courteous driving style than if they employed more selfish behavior.

I find it fascinating that, as humans, we rarely think about our driving behavior, whereas we have to explicitly tell a robot car how to behave and interact with others. When programming robots, we need to make so many choices about their behavior that it would seem like an insurmountable task—even if we knew what humans would decide in all sorts of circumstances. But here's the thing: our knowledge of human behavior is imperfect at best and dangerously misleading at worst. When you read the example about courteously changing lanes, do you feel that it represents your own behavior? I'm sure some of you are enthusiastically nodding right now, while others probably think that the example is completely unrealistic. This is because driving is about making decisions—about communicating and interacting with others in complex ways. And as I will show you throughout this book, these kinds of behaviors can strongly differ across countries and cultures.

Consider, for a moment, what would happen if you tried to hail a US robotaxi in Cairo, Egypt. The car arrives, you get in, and . . . you wait. The turn signal indicates that the robot car wants to merge into the quickly passing traffic, but it isn't moving. Five minutes later, it is still standing where it picked you up, waiting for a gap in traffic that it could merge into. But that gap never arrives. Eventually, you give up and exit the car, but not until after you leave a scathing review of the incapable "driver" that you never even met.

What do you think went wrong in this scenario? Clearly, the robot car was trying to enter the traffic and get you to your destination of choice. But its strategies—waiting for a gap in traffic and hoping someone sees the turn signal and mercifully lets the car in—didn't work in this context. This is because Egypt, like so many countries across the world, is an interesting place for drivers. The driving behavior of most Egyptians can't be defined in terms of courteous versus selfish, as the team of Berkeley researchers observed in the US. Instead, many Egyptians drive in a way that Americans may describe as assertive and perhaps even reckless at times. Christian Sturm, a collaborator of mine who has spent several years living in Egypt, once told me that there is an art to driving in Egypt, one that requires understanding the dynamic and unwritten rules that orchestrate the organized chaos. To change lanes or reenter traffic, he said you just need to loudly honk the horn— make sure to hold it for a second or two so people don't think you're just saying "hi"—and then just go for it. Maybe emphasize your intention by waving your hand out of the window. All else will follow.

You can see how robot cars in Egypt would need a serious collision damage waiver—if they're able to enter the traffic at all. More likely than not, a cautious and defensive driving style that works so well to prevent accidents in the US would completely paralyze them in Egypt and alienate anyone involved. And it's not just about being too defensive. Robot cars also don't know other local, unspoken traffic rules, such as that briefly honking is meant to be a greeting, that women with headscarves should get the right-of-way, and that pedestrians can suddenly appear on the road at any moment no matter where. Without the robot cars adapting to local norms, people would almost definitely become frustrated and lose trust in the technology; they may experience an increase in accidents, or they may be forced to adapt to the robot cars' practices and norms to avoid them. Writing about autonomous driving in the Arab region, the researchers Hesham Eraqi and Ibrahim Sobh described the root of the problem: "Unfortunately, the overwhelming majority of the data collection efforts worldwide are not recorded in cities with chaotic driving."[73] In their article, they list several "unconventional" traffic situations that are nevertheless common in Egypt and beyond, such as the sharing of roads with tuk-tuks (a kind of tricycle used to ferry people around) and donkey carts, missing lane markings, passengers on the back of trucks (which robot cars could mistake for pedestrians), drivers driving on the wrong side of the road, vehicles parked in unauthorized areas, and so on. The vast majority of these situations would be unknown to robot cars that were trained on data collected in a very small part of the US.

There are several insights in this story that showcase what this book is about. One is that it is an example of what I call *Digital Culture Shock.* I use this phrase to describe the experience and influence of actively or passively using technology that is not in line with one's cultural practices and norms. Just like culture shock in interpersonal interactions, the experience of interacting with technology that behaves unusually to us can be eye-opening and make us more adaptable. But it can also lead to misunderstandings, leave people feeling alienated, negatively impact their trust, and force them to adapt to the technology's practices and norms. *Digital Culture Shock* analyzes this power that technology creators have to influence societies and their cultures.

The robot car story also demonstrates that technology is inherently cultural; it works in a specific cultural setting but often fails in others. When we let robot cars drive around to collect data on their surroundings, they learn, over time, to become better at functioning in a specific environment. We're essentially training them to become more culturally adept in a specific local context. The fact that the Waymo robot cars were exclusively trained on traffic situations encountered in large, urban centers in the US means that they are going to be overwhelmed by street environments and traffic situations in parts of the world that function differently. I would predict that a car with roots in the US will find it difficult, if not impossible, to navigate many other places—places where the robot cars haven't learned the unspoken rules about who gets to go first, how one overtakes other cars, and what one, two, or three short honks or one long honk mean, or where there is simply a little bit more unpredictability. So another insight here is that technology can be less useful and intuitive—and even dangerous—when it gets transplanted from one local context to another. (Tech optimists may argue that robot cars could be easily retrained to work in other countries. I would counter that many places lack the infrastructure to let robot cars collect the necessary daily terabytes of data, and support this collection for 10+ years. And even if this infrastructure were in place, there are still many assumptions built into autonomous cars that would not apply in most parts of the world.)

The point is, just as robot cars are trained on data that they gather in specific cultural settings, so is all of the digital technology we use. Technology designers and developers come with their own data in the form of cultural values and norms, which they subconsciously embed in any design decision, from the initial idea for a product to the final button. The core assumption of robotaxis is that there is a need for transporting a single passenger from A to B. This assumption probably originated from the developers' experiences in the US, where over 76% of people are driving solo to work[275]—but it doesn't apply in places where people most often take minibuses or share a car with multiple people.[136] What this means is that by deploying technology in places with different cultural practices, technology creators are imposing their assumptions on others. And, with that, they wield enormous power to change not just individual users, but whole societies.

The WEIRD Problem

After dedicating my research career to studying how digital technology is designed and used by people around the world, I have become convinced that it is almost never one-size-fits-all. What someone in one place of the world dreams up as being beneficial and intuitively usable is often misaligned with how people in other places perceive it. This is because humans are immensely variable in how they live their lives and how they may use technology to support them.

The examples of robotaxis and ridesharing can illustrate some of these differences—and what can go wrong if one ignores them—but they still leave open a key question: Why do technology companies so often ignore the myriad of cultural differences when they deploy technology across the world? One reason is that technology companies are primarily interested in capitalistic gain—money rules the technology world and, unfortunately, that means the focus is not on optimizing websites, apps, robot cars, or other products for all its users (and perhaps creating different versions through a costly process), but on optimizing it for those that they can make most profit from. One-size-fits-all is always cheaper. But after researching this topic for almost two decades, I have come to realize that there is another issue at play. It's the issue of knowing how to create technology that is better suited for people who are unlike ourselves. Even if technology creators have learned about human behaviors, needs, and preferences, what constitutes good design, how people perceive information, or how they interact with technology—knowledge that is still rarely taught in even the top computer science programs—they often don't realize that most of these assumptions don't generalize across countries and cultures. This is because the vast majority of what is known about people is based on scientific studies with North American undergraduate students. We simply know most about people in the US, and even there only about a tiny slice of its very diverse population. In 2008, psychologist Jeffrey Arnett published an article titled "The neglected 95%," referring to his finding that studies included in premier psychology journals primarily report on American participant samples,[19] most of them college students who participated in the studies for course credit or a small financial reward. According to his analysis, only 3% of the participant samples were in Asia, 1% were in Latin America and Israel, and less than 1% were in Africa and the Middle East—despite the vast majority of the human population living in these parts of the world. When the article came out, it sent tiny shockwaves through the psychology community, though I'm sure nobody was entirely surprised to see these numbers. As Arnett describes in his article, psychology research has always tried to be like the natural sciences. As with the laws of physics, which we would expect to apply anywhere in the world, psychologists have been trying to understand basic human behaviors by stripping away any unruly variables of real life. "The goal was to identify human universals, the fundamental processes and principles that comprise human psychological functioning," Arnett wrote. And often this meant that scientists assumed their findings were indeed broadly applicable. Even today, technology practitioners readily incorporate this "scientifically proven" knowledge into their designs.

But humans are complicated. Whatever behaviors scientists observe in their studies may not generalize to other people and contexts. Scientists call this a lack of *external validity*, which is a fancy way of saying that a single study rarely holds all the truth. Change the context, the people, or something else about the study, and you may have completely different results. The findings cannot be generalized. And this is more common than you'd think. In an influential article published in 2010 in the journal *Behavioral and Brain Sciences*, psychologists Joseph Henrich, Steven Heine, and Ara Norenzayan showed that much of what we've assumed are universal traits across humanity—from visual perception and spatial reasoning to self-concepts and moral reasoning—can be debunked as findings that do not apply to people in other parts

of the world.[122] The main issue, they wrote, is that common participant samples are all too often Western, Educated, Industrialized, Rich,
and Democratic, or in short, WEIRD. And these WEIRD samples
are often very unusual in how they perceive information, reason, and
behave compared to people in other parts of the world. Americans, the
research team concluded, might even be "the WEIRDest people in the
world."[122,*]

Unsurprisingly, WEIRD people are not representative of the values
and ethical principles that govern societies around the world. To stick
with the topic of self-driving cars for a moment, think about what would
be the morally right thing to do in a given traffic situation. What should
a robot car decide to do when a crash seems inevitable? In a *Nature*
paper published in 2018, a group of researchers led by Edmond Awad
described their endeavor to find the ethical principles that should be
programmed into robot cars.[21] Rather than leaving these decisions to
car manufacturers and tech companies, AI ethics, they reasoned, needs
to involve the public for robot cars to be accepted and to be in line
with social expectations. They set out to create the Moral Machine,
a website that asks people to decide on a number of moral dilemmas
that a self-driving car may encounter. Faced with an imminent crash,
should a robot car hit the humans on the road or swerve to hit a barrier,
thus killing the passengers? What if it swerved and killed pets rather
than humans? Should it decide to save the young or the elderly? People with lower or higher social status? And so on. The choices seem

---

*Henrich later argued in his book with the same name[121] that Western people's psychology
changed due to the medieval Catholic Church prescribing rules about family and marriage. For
example, he described how the ban of cousin marriage to some extent abolished the many inter-
connected clans and weakened kinship ties, enlarging the influence of independent individuals
and families, whose success increasingly relied on differentiating and standing up for themselves. Individualism was rising. Several people later criticized this theory: It may explain why
we see cultural differences across the world, they argued, but the cultural evolution that Henrich described could also be used to claim "that there is something 'superior' about Western
culture."[117] For example, Nicholas Guyatt wrote in a *Guardian* book review: "Henrich might
wince at the suggestion that *The Weirdest People in the World* endorses social Darwinism, but in
its emphasis on the supposedly discrete nature of culture and on the virtues of 'weird' thinking
and progress it comes uncomfortably close to doing just that."[109]

gutwrenchingly cruel, but it turns out that when forced to decide, people often have consistent opinions on what is right and what is wrong. The only thing is . . . there is no such thing as a worldwide consensus on ethical principles. Among the responses from 130 countries with at least 100 participants, the research team found three main clusters of countries: In the Western cluster (including the US and several European countries), people clearly prefer to let fate run its course and not intervene—though if forced to decide, they would rather spare the young and people who are fit. In the Eastern cluster (including East Asian and Islamic countries), people prefer to spare pedestrians over passengers and the lawful over those jaywalking. And finally, in the Southern cluster (including Central and South American countries), people would rather save the young over elderly people, those with higher social status over those with lower social status, and females over males. We can see from these findings that there are *some* ethical principles that appear to be universal, such as a preference for sparing human lives over pets'. But as the authors argue, there are also several preferences that vary across countries, many of which can be attributed to cultural differences in acceptance of inequalities in society.

The Moral Machine has taught us that programming robot cars to decide in these rare but tricky situations requires more than a bit of thinking about how we would want the car to behave. What we need to program these machines is detailed knowledge about what people believe is right. About their moral stance and how it may differ across the world. This is what makes studies like the Moral Machine so necessary, because they not only show how individual preferences can vary, but also how they may have been influenced by culture.

Alas, nearly all research studies look at much smaller numbers of participants, which sometimes means that it is simply unknown whether findings generalize. And because most studies focus on WEIRD participants, the data one can work with represents only a small fraction of the world's population. This is not just a problem in psychology research, but really in all fields that investigate humans or things humans interact with. It seems obvious, but hear me out: One should never develop technology in a vacuum without thinking about how humans will interact with it. My research community, in the field of Human-Computer Interaction (or, for short, HCI), is passionate about exactly that. "How do people interact with technology?" and "How can we design technology to better suit its human users?" are key questions HCI researchers try to answer. But while the field has contributed to really important advances in how we use computing technology today—such as how we swipe on a mobile screen, why people are susceptible to online misinformation, or what makes technology usable— it has also had a strong WEIRD bias. As my collaborators and I found, 73% of the studies that were published over the past few years in one of the research community's premier venues, *CHI*, exclusively looked at Western participants, and almost all of them were educated, industrialized, relatively rich, and democratic.[165] When my collaborators and I had finished analyzing who is being studied, we were stunned. Unsurprisingly, the vast majority of studies were conducted in North America, in line with the US dominating the number of researchers in the field. We expected that. But more than half of the world's countries, 102 countries mostly in Africa and Asia, had not been studied at all during the past five years. What we know about human-computer interaction and how we develop and test technology is even WEIRDer than my collaborators and I had assumed.

Designing for Cultural Diversity

A lot of this book is inspired by these findings. Humans are generally more similar than different, but variation is the norm. If people aren't universal, how do they design, use, and perceive technology differently? What happens if people from one culture create technology and those from another use it? And what can technology design learn from our world's cultural diversity? Throughout the book, I will answer these questions by shining a high-lumen spotlight on research that has questioned human universality and studied less WEIRD people.

In the first six chapters, I draw on decades of research in anthropology, psychology, neuroscience, and human-computer interaction to establish that people's norms and behaviors vary in ways that can directly affect their use and perception of digital technology and its user interfaces. I combine insights from very different methodological traditions within the field of human-computer interaction, including ethnographies, interview studies, surveys, and quantitative laboratory and online research. A substantial part of these insights are based on the cross-cultural research I have conducted with my PhD and post-doc advisors, students, and collaborators. I start by showing how digital technology is like a kaleidoscope of cultures: Its design choices are plentiful, consisting of innumerous colors and viewpoints that can be put together in an endless number of ways. Looking at how technology is designed across the world can help us understand how one size does not fit all and unlock the creativity needed to think beyond what we are used to thinking. In Chapter 3 I show that rather than our being born with a certain cultural mindset, new experiences continuously shape our own culture, which in turn—through interaction with others—shapes those around us and our society. Our malleable brains play an important role in this cultural cycle by constantly optimizing their cognitive performance so that we can best deal with any incoming information. We continuously learn how to predict the norms around us. This brain plasticity, as neuroscientists call it, also allows us to code-switch, be multicultural, adopt new cultural traits, or even be temporarily primed to think more like a person with another cultural background. It's a fascinating world that we live in, in which the cultural diversity can open our eyes to how we adapt to our environments.

Technology is yet another actor in this cultural cycle. It subtly conveys the values that were subconsciously embedded in its design by its creators. When technology developers and designers decide to create a technology product that addresses a certain need, to include functionality and guidelines that align with their values and norms, or to let it learn from a biased set of datasets, they invariably include cultural beliefs that are not universally shared by people around the world. There is simply no one way to design technology. As many people have said before me, technology is never value- or culturally neutral.[82–84,304,305]

In my eyes, there is almost no right or wrong in terms of what "culture" a technology should have (though most people, myself included, endorse specific values more than others), but these decisions do have an enormous impact. While the first few chapters of this book will show you how technology design can augment people's lives—for example, by helping them be more efficient, protect their security and privacy, intuitively search for information, effortlessly navigate nonlinear content, and increase their trust and feeling of belonging—the later chapters in the book will explore the more insidious effects of digital culture shock. Chapter 8 is the heart of this book because it discusses what happens if technology isn't designed for you. I argue that people can experience symptoms similar to culture shock in interpersonal interactions if there is a mismatch between users' values and norms and those imprinted in a technology product's design. The technology may not meet our expectations: perhaps we occasionally become confused by its design, don't find it intuitive to navigate, don't perceive ourselves as belonging, or are hit by assumptions and stereotypes that surface in its design. I anticipate this chapter will provoke technology developers and designers who may not have realized the power they have over their users and whole societies. You might even feel a bit *culture shocked* reading about all of this—after all, experiencing new ideas and viewpoints can make people feel uncomfortable. But feeling culture shocked can also be a method for engaging and wrestling with issues new to us, and I'm grateful if you are open to experiencing this. Chapter 9 takes these provocations further by proposing that technology design decisions can be perceived as marginalizing and oppressive by those who are negatively affected by them. I discuss how it is a pervasive myth that technology products are innocent tools that are voluntarily used by people around the world, and contrast this with research that has likened it to postcolonialism, digital neocolonialism, and cultural imperialism. If you think that these analogies are too gloomy, then I hope this book will succeed in convincing you that everything depends on your viewpoint: For those who have been on the receiving end of technology that clashed with their cultural values and norms, feeling alienated, marginalized, and oppressed by technology is the reality.

Those whose values have so far mostly aligned with the digital technology they've had in their lives may find it difficult to relate. If you fall into this category, I urge you to imagine the roles are reversed.

Chapter 10 presents a path forward. I describe different approaches to how developers and designers can resist culture shock in technology by designing not just *for* users, but *with* users, including approaches such as participatory design, postcolonial computing, and decolonizing design. None of these make for easy fixes. The approaches take more time and engagement than most companies will be willing to invest in a capitalist world. Hence, this chapter will also walk you through some of my prior work with industry, through which I have become convinced that we need to find compromises between the ideals of academia and the realities of industry. Both sides need to develop approaches that are pragmatic enough to be implemented on a regular basis and ambitious enough to cater to their diverse users, including mitigating forms of cultural imperialism and neocolonialism. The reason for engaging with design approaches suggested by academia, I argue, is that access to technology is a social justice issue: It is a moral obligation to treat everyone the same and to not disadvantage groups of people who may be unlike ourselves. Technology companies should go the extra length to find out how the technology products that they make money with might negatively impact people in this world.

Developing culturally aware technology is of course not as simple as that. Producing technology that really fits all will require disentangling a tightly tangled ball of national interests, historical influences, policy decisions, capitalistic choices, and corporate strategies. After all, tech companies and governments are driven by power in the form of influence and money. I will lay out that it will require using both carrots and sticks to hold technology providers accountable for any harm that the common one-size-fits-all designs cause. The stick could be writing into law that companies need to provide reasonable adjustments for people from diverse countries and cultures, similarly to how many countries require them to not discriminate against people with disabilities. The carrot is perhaps more straightforward: As I argue throughout this book, there are enormous opportunities in the cultural diversity in this world, including in the cultural diversity of technology. We should all work to understand and preserve it. I see this cultural diversity as a boost to our creativity and as an opportunity for rethinking how we could design the technology of the future. If companies want to remain competitive, unlocking new and creatively designed technology products could be as simple as learning about the diversity of cultural norms and behaviors in this world.

Since I will be trying to cram human diversity into relatively few pages, let me dive a bit more into what I mean when I talk about culture. I don't mean it to be the same as a country or even a continent, because there can be much variability across people within geographical and political borders, and it would also ignore cultural groups that span multiple regions.[183] And I don't mean it to be an individual phenomenon, because, as most people would probably agree, I also believe there are many behaviors and tendencies that are shared by groups of people. Instead, the way I'm using culture in this book is to describe social groups that share learned values, norms, and behaviors.[46] Not everyone in a group will behave the same way, but if you took a representative sample of that group, you'll likely see that there are common values and actions. Sometimes these are national groups, such as people from the same country, though we'll also discuss how language, religion, race, and other factors can shape subcultural groups within a nation, and how artificial country borders (introduced by external forces such as colonialism) can mean that parts of the population may be more similar to people in another country than to people in their own country.

All of this is terrifyingly complicated. Culture simply can't be well defined, which is already noticeable in the fact that it is often described as a set of shared values, norms, and behaviors. There are also anthropologists who would rather get rid of the term "culture" altogether because it seems to imply that it is some kind of tangible and stable thing, whereas culture is actually extremely variable and dynamic.[16] In addition, some argue that culture is a tool for "othering"—for exaggerating the differences between an anthropologist and the people they study and for placing them in a subordinate position.[3] Yet many researchers have also suggested that the term itself is useful, as long as it is used "to its best intents."[46] I tend to agree that there is something to it. As inadequate of a concept as it may be, people refer to the term "culture" all the time when traveling abroad, meeting newcomers in their country, outsourcing production, or talking to others online. Why do we find the term "culture" so useful if it's so difficult to describe? I think it's because labels add boundaries to an infinitely complex world. Contrasting one culture with another, for example, helps us categorize behaviors and norms and lets us better understand our own identity. And knowing about culture—being able to label it—also helps us explain situations in which we experience miscommunication or even feelings of culture shock.

But using labels in the context of culture doesn't come without risks. Anytime we label things, such as when I refer to "Egyptian drivers," or "the Germans," it risks oversimplifying, stereotyping, or even involuntarily creating the notion that one side is better than the other. Adam Alter, a marketing professor at NYU's Stern School of Business, put it well in his book *Drunk Tank Pink*: "Social labels aren't born dangerous. There's nothing inherently problematic about labeling a person 'right-handed' or 'black' or 'working class,' but those labels are harmful to the extent that they become associated with meaningful character traits."[13, p.34] Indeed, labels that we merely use to make our complex worlds simpler often become generalized and loaded with meaning. Labels can suddenly imply that people hyperfocus on specific associations between a label and specific character traits while overlooking other, potentially more positive characteristics. To give you an example, did you know that simply seeing a child play in a wealthy neighborhood will likely make you think of the kid as smarter than one playing in a less wealthy neighborhood?[64] Or that being labeled as a "bloomer," someone who is expected to show greater intellectual growth, could increase your IQ score by 10–15 points in subsequent tests?[232] It's because humans tend to connect labels with specific character traits, as Alter points out. When we see a child play in a wealthy neighborhood, we subconsciously apply labels like "wealth," "rich neighborhood," "educated parents"—and then use any correct answer the child gives us as proof to back up our initial impression. Any incorrect answer we will happily overlook and discard as an outlier. Similarly, when teachers think of you as a bloomer, they will label you as smart and ambitious. Each time you answer a question correctly, your teachers will use it as further proof that you are highly intelligent. Each time you get it wrong, they'll think of it as an anomaly, and maybe try even harder to get you through your blooming phase.

So labels have to be used and interpreted with great care. As Alter describes, knowing that they exist can help us take advantage of them when they're helpful, such as for understanding the diversity of people in this world. But we need to resist them when they hurt and actively work against our human tendency to overgeneralize. This also means that any study findings I report on in this book have to be seen as just that—as findings that apply to the particular participant group that was being studied.

For simplicity, you'll often see me write statements like "Germans schedule events 22 days earlier, on average, than people in Lebanon," which only applies to the context of the specific study I'm describing. In this case, it is true that the average German user of the event-scheduling tool Doodle scheduled events 22 days earlier than the average Lebanese Doodle user. But of course that doesn't mean you'd see this behavior in all Germans—Doodle users may be particularly into advance planning, for example, and they are also younger than the average German— and the "average" behavior, by definition, means that many people will schedule earlier and later. So there is much variation in the data and the findings may not generalize to other people outside of the sample that was being studied.

There is another important detail you'll have to add to any simplifying sentences, and that is *when* the data was collected that each study reports on. People's values and behaviors change all the time, which makes culture a constantly changing, dynamic phenomenon. Psychologists and anthropologists describe it as a cultural cycle in which everyone plays a role in shaping and receiving dominant values, norms, and behaviors.[114,179] As a traveler, as a colleague interacting with others at work, or as an expat in another country, you too are subtly influencing other people's cultures and invariably adopting parts of theirs. But it's not just humans who are playing a role in this; it's also artifacts, food, really anything that travels around the world. Technology plays a huge part in this too, and not just the memes that become viral sensations on the Internet. In fact, I think there is much to be said about computing technology being yet another social actor in this cultural cycle. All technology implicitly encodes social and cultural values, which it can instill in others in this ever-evolving cultural cycle.

Throughout the book, I'll try to connect the various ways of using and designing technology to cultural theories. These theories were created by anthropologists like Triandis, Hall, Hofstede, Trompenaar, Schwartz, Inglehart, Pelto, and many others to understand cultural differences. (And yes, these researchers and theories, too, are very WEIRD.) You may have heard about *cultural dimensions*—essentially, these are labels that anthropologists have given different groups to compare them to others. One of the most important ones is individualism, used by the French aristocrat, diplomat, and historian Alexis de Tocqueville to describe his insights from a visit to America in 1831. In his book *Democracy in America*, Tocqueville wrote about the democratic American spirit he observed during his travels, with its comparatively high self-reliance, personal freedom, independence, and equality. According to him, Americans think of themselves as separate individuals who hold their destiny in their own hands.[273] Several anthropologists have assigned similar characteristics to the term "individualism," often contrasting it with cultures of collectivism, in which people are thought to be less focused on the individual self but tend to emphasize their duties, roles, and responsibilities toward an ingroup, such as their family.

If you've heard of individualism versus collectivism, you may have immediately thought of Geert Hofstede. He is one of the most widely cited social scientists, and probably most famous for coming up with a set of cultural dimensions in 1980, one of which is individualism/collectivism. What Hofstede and many other anthropologists did to find these cultural dimensions is actually quite remarkable when considering that culture is usually thought of as inherently intangible: They deployed surveys in various different countries and, through a complicated process of statistical analyses, found which dimensions are most relevant to describing differences in people's values and norms. In many cases, the outcome was a set of scores that tell us where each cultural group sits on a particular dimension—usually national cultural groups, though some anthropologists have also looked beyond country borders. These dimensions and scores give us a way to compare and reason about cultural differences. Often, cultural dimensions have helped researchers justify why they are comparing specific countries or why they are seeing certain phenomena. Many of the scores seem to intuitively make sense; for example, Hofstede's scores squarely rank the US as the most individualistic nation of the 40 countries he analyzed,[124] which seemingly confirms Tocqueville's observations. But the cultural dimension scores have also been heavily criticized[212] and are frequently revised.[265] So while the labels that cultural dimensions assign to people, cultures, and countries have been seen as useful for explaining similarities and differences researchers observe, we need to take great care to avoid overinterpreting them. Humanity is simply more complex than any set of cultural dimensions and scores could ever describe.

Here's one more word of caution. There are many other cultural theories, dimensions, and research findings that I'll refer to throughout the book, but these, too, come with their own downsides. For one, most theories are well substantiated based on what we know today, but that doesn't mean they couldn't be debunked in the future. This is just as when humanity came to realize that the universe is constantly expanding, which led us to suddenly notice that Einstein's theory of a static universe wasn't quite right after all. The second reason to carefully interpret these theories is that the studies leading to them often have the same sampling issues as most other human-subject studies. Hofstede's cultural dimensions, for instance, have been criticized because they are based on survey responses by IBM employees in different countries.[183] These people may be a very particular subset of these countries and they may be highly influenced by IBM's *corporate culture,* just as undergraduate students are not representative of the general population. Hofstede used these surveys to explain differences in work-related values, but these differences have often been interpreted to be comparisons of *national culture.* And then there is the issue of the cultural cycle I was referring to above. If culture is constantly changing, can we still use the cultural dimension scores years after they've been established? Well, let's just say it's debatable. Many people would say that Hofstede's scores are now outdated, though they have been updated to some extent[269] and many researchers continue to find them useful. His dimensions have also been found to correlate with dimensions other anthropologists have come up with, such as the cultural dimensions that came out of the World Values Survey—a large-scale survey that has been repeated in seven iterations since 1981.[129] So altogether, cultural difference can be often explained using these dimensions, but it is important to know their limitations.

Many times throughout this book, I will compare technology designs developed in Western countries with those in the rest of the world. There are several reasons for this: One is that four of the five countries I have lived in are considered Western countries—the only exception is Rwanda. I was born and grew up in a Western country, Germany, I attended high school in Australia, I did my PhD in Switzerland, and I currently live and work in the US. Given these influences, my knowledge of the technology world is naturally skewed. It also doesn't help that most cross-cultural research (at least what has been published in English) has been done at US-based institutions, which means that there's been an implicit bias to use the US and its people or digital technology as the baseline, the "norm" to compare against. Another reason is that according to the CB Insights' Global Tech Hub Report, the US, and particularly Silicon Valley, still leads the list of technology hubs in the world in terms of financial investments.[51] For the past decades, the reality has been that Silicon Valley has developed many of the digital technologies that are globally used. The rest of the world has just had to live with it, without local alternatives in many cases. Fortunately, this is changing quite rapidly. There are now many technology hubs across the world that are seeing large investments in startups and that have had international successes with several technology products. A few of the larger hubs and their successes are in London (fintech company TransferWise, food-delivery company Deliveroo), Seoul (Coupang), Tel Aviv (Waze), Stockholm (Spotify), Bengaluru (Olacabs), Mumbai (Housing.com), New Delhi (Snapdeal), Tokyo (Mercari), Sao Paolo (mobile gaming company Wildlife Studio), and Australia (Canva). Seeing how they do things differently, and how the technology coming out of these tech hubs serves different needs with different designs, is a joy for me to unpack and something that I hope we can all learn from.

And then there is China. The country has some of the largest tech hubs in terms of financial investment, heavily driven by the Chinese government investing in technology innovations and by its large population that provides an eager market for new tech. Beijing is the second-largest hub in the world, bringing in 72 billion US dollars between 2012 and 2018 compared to 140 billion US dollars invested in Silicon Valley companies.[51] And while they may be a distant second to Silicon Valley right now, Chinese technology hubs, like Beijing and Shenzhen, are home to an increasingly large number of technology companies, several of them global. For example, WeChat is the Chinese equivalent to Facebook or Twitter, but provides many features that address China's unique political context.[295] JingDong (JD.com) is an e-commerce giant that is now operating in Brazil, Cambodia, Chile, China, Denmark, Ecuador, Finland, Indonesia, Laos, Norway, Peru, Sweden, and Switzerland.[307] TikTok, the video-based social media platform invented by the Chinese company ByteDance, has become a global sensation (though it has retained a Chinese version, Douyin, which has its own content ecosystem for the Chinese market). And Chinese AI technology company Sense-Time, which offers a vast array of technologies that other companies integrate in their products—from face and image recognition to video analysis and autonomous driving systems—operates in Hong

Kong, Mainland China, Taiwan, Macau, Japan, Singapore, Saudi Arabia, the United Arab Emirates, Malaysia, South Korea, and several other places. As Silvia Lindtner, professor at the University of Michigan and author of *Prototype Nation: China and the Contested Promise of Innovation*[164] told me, Chinese entrepreneurs and technologists are increasingly breaking away from their image as copycats. They are creative players in hardware and software, innovating at incredible speed and supported by a readily available market of technology adopters.

Needless to say, there are lots of politics at play. Technology supremacy promises a tremendous amount of power over other parts of the world. But while political leaders are most often concerned about technology's economic power and data privacy issues, I think an imbalance in who creates technology is problematic for additional reasons: because technology designs implicitly embed cultural values, which often makes them better suited for people in one part of the world than for others; because technology has the power to change our values over time, which has been likened to imperialistic and colonialist powers and could potentially disrupt the social glue that holds a society together; and because technology can cause us to experience culture shock, which can make us more world-open and adaptable, but can also have disturbing negative effects.

And this is ultimately what this book is about. It's about questioning the assumptions that one intuitively makes when designing technology, and about giving you the tools to detect whether you are developing and using technology that is misaligned with your culture.

One goal I have for this book is to help readers see that people may design, use, or react to technology quite differently, and that this is because we've all had different cultural experiences that have shaped our behaviors, values, and even our cognition. I hope you will see this when just reading the book; but for the avid reader, I've included several online tests here and there. These tests let you participate in short experiments in a virtual lab, the LabintheWild, and get personalized results that tell you how you compare to others. I developed

LabintheWild in 2012 to study how people use and perceive technology around the world,[227] and my lab and others have since run over 100 studies on it with people from more than 200 countries and territories. If you participate, we will release your data (in a completely anonymized form) so that researchers can use it to make science a little bit less WEIRD. Resist interpreting too much into the labels that the personalized results may give you, but embrace them when you think they help you understand where you're coming from when designing or using technology.

2

Kaleidoscope of Cultures

It's been more than a quarter of a century since Google started taking over the majority of the search market share in the US and later in many other countries. What its homepage offered was . . . well, not much. Just a plain white background, the Google logo, and a simple text box waiting for people to type in whatever was on their minds. It's as if Google felt like it needed to put blinders on people so that they could focus on searching the Internet without any distractions. Until today, it remains the web page with the lowest word and image count I know. Google has been lauded for its no-frills user interface, which has contributed to its success.

I can't deny that I like Google's approach to design. It perfectly corresponds to my idea of good user experience design: It's easy and efficient to use and meets my needs without imposing unnecessary functionalities that would distract me from my goals. So I've never been surprised that Google has dominated the search engine market for so long. Except that it hasn't, at least not in all countries.

In 2007 I visited South Korea for the first time and learned that South Koreans don't "google"—they "naver." In fact, in South Korea, Naver.com has been fending off Google's quest for world domination for the longest time. And with much success. In May 2023, Naver held more than 50% of the South Korean search market share according to some sources.[48] My curiosity was piqued. Why do so many South Koreans prefer Naver over Google? One answer is that Naver approaches search entirely differently from Google. Have a look at its user interface in Figure 2.1, or, better yet, navigate your browser to Naver.com where you can see blinking animations and colors in all their beauty. Even if you can't read Korean, you will quickly see that Naver isn't just a website to search the Web. Its web portal–like interface combines almost everything a person may need to do on the Internet. Think of being able to search, know the weather and latest stock market results, shop, browse through questions and answers by others in the community, pay your friends, book flights, and much more—all from one single access point. You would never have to leave the page.

One of Naver's defining features comes from its need to provide search results in Korean, which means relying less on English web pages in its search results. With its *Knowledge iN* service, Naver started letting users ask and answer questions.[194] The approach is more similar to a Question & Answer platform than to a search engine, but it has a huge benefit: Rather than getting search results from unknown sources from all over the Web, Naver's *Knowledge iN* got results from a community of people. Today, Naver offers what one could call "comprehensive search," showing results in a single page gathered from a combination of external web links and user-generated content. This makes it a one-stop shop akin to a local corner store where you can get trusted insider information in addition to world news. It's the perfect place on the Internet for a collectivistic society that is more likely to trust information provided by an ingroup—a group of people that one feels part of—than to trust information found in the outgroup corners of the Web.

Of course, Naver's popularity in South Korea didn't go unnoticed. Google had already aggressively expanded its business to many other countries, but it was somewhat stuck in East Asia. Dan Russell, who led Google's Search Quality and User Happiness division as a senior research scientist for many years, once told me that Google's market volume was so low in South Korea that its Korean design team was given carte blanche to do whatever they thought would work. They started experimenting with making the South Korean search page much more playful, which Russell thought was more in keeping with Korean user interfaces. "Let animated GIFs abound!" he told me about one of

(a) Naver

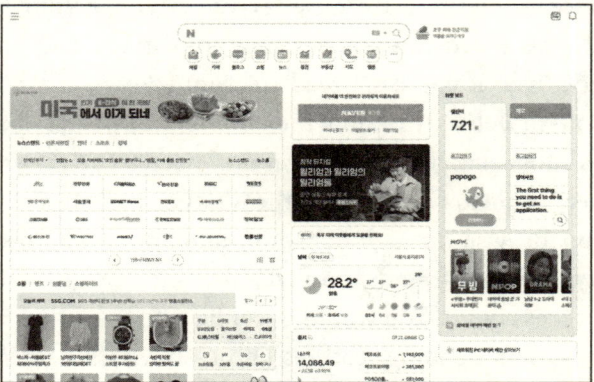

(b) Yahoo! Japan

(c) Google

FIGURE 2.1. South Korea's search engine Naver.com, Yahoo! Japan, and Google in comparison.

the attempts the designers made. The Korean designers created more playful versions of the Google homepage, focused on adapting the aesthetic appeal of the page while still staying true to the Google brand. "But it never really seemed to do much to overall usage patterns. And it was never clear why," Russell later said.[236]

Google had no idea what was going on in places like South Korea, Japan, and China. While its search engine quickly took off in many other countries, Google employees were tearing their hair out in East Asia. Yahoo! Japan, a Japanese Internet company that was jointly founded by the American Yahoo! and the Japanese SoftBank company, was long in the lead (and still holds a large chunk of the market share in Japan) after heavily investing in localized content and integrating its search engine with multiple services, such as shopping, finance, music, and more.[140] In line with many other Japanese web pages, Yahoo! Japan's homepage is filled with information and services, organized in several panes and overlaid in tabs. (As we will see in Chapter 3, Japanese people tend to be exposed to more visually complex environments, including websites.) In China, local rival Baidu, which has a look not too different from Google's minimalist interface, has secured its dominance, though the fact that Google is regularly blocked by the government may be the main reason here.[26]

In general, there isn't just one reason for Google's difficulties in gaining market share in these countries. It is a myriad of cultural and political particularities that to this date pose a challenge for Google. In the case of South Korea and Japan, I think Google's approach to search is perhaps a little too Western. You see, the Google search engine and a few of its Western-designed predecessors (remember Lycos, anyone?) have defined search to the extent that many people now believe they can apply the same search behaviors to any information system out there and get reasonable results. Type in a few keywords and relevant results will magically appear. Initially, typing in a chain of query words was indeed the only way one could efficiently search Google. This is because Google's developers thought this is the way people can easily search: using taxonomic terms. The words "broken bike chain fix" that I frantically typed into the search box this morning implicitly convey the question, "How can I fix my bike chain that just broke?" (Yes, this morning could have gone better!) But as Dan Russell points out in his book *The Joy of Search*,[237] many people like to search by typing in full questions. When search engines got better at interpreting such questions, people started using them more. As a former Google researcher, Russell has probably seen gazillions of search queries (though maybe not quite a googol), and while many are fairly standard, he has found that there is still a lot of diversity in how people search. This is also the reason why Google's Senior Vice President Prabhakar Raghavan suggested in 2022 that the company will set up local teams in several Asian countries because—wait for it—Google needs to adapt its services to other countries and cultures.[267] What the company had noticed is that there are not just individual differences in how people search, but that their search approaches strongly vary across countries. Raghavan noted that Japanese people tend to be very succinct, typing in a few keywords only. Indians are the exact opposite, with around a third of the search entries being spoken queries that tend to be much longer. Google's usual approach of getting people to adapt their search to the way that its search engine was designed doesn't quite work in these countries. Instead of processing a few search keywords, Raghavan decided that the company needs to work on better understanding spoken language queries in the 22 official Indian languages.

I had all sorts of emotional reactions when I first heard about this. For one, I was relieved. "Great that these users will finally have a version of the search engine that works for them!" I thought to myself. But also: "Wow, it's 2022 and you're telling me that Google needs to adapt its search to other countries and cultures?" The thing is, researchers in my field, human-computer interaction, have preached the need for understanding users for several decades. We call it "user-centered design" if one tries to deeply understand the user, their needs and habits, and how they will interact with a product. It's like a mantra we keep repeating to ourselves, to our students, really everyone. Ideally, this mantra is followed throughout the development process, from the initial idea all the way to the implementation of the final product. And ideally, it takes into account human diversity. Google practices user-centered design in many ways, but like most (all?) companies, its motives are ultimately shaped by capitalist forces. The markets that have been of primary importance to Google are the US and Western Europe—this is where they thought most of the advertising revenue was to be made. So Google invested in understanding these markets. But while a user-centered design approach would suggest understanding users *before* developing the product, I'd say that most of Google's user insights are actually gained *in hindsight* by observing how people search. Naturally, this is a great way to learn about people who already use your product, but not so much to learn about nonusers.

Indeed, Dan Russell confided in me that most of the search experiments he used to run at Google involved large numbers of users that would make most researchers turn green with envy—but they were primarily run with Google users in the US and Western Europe.[235] It is these WEIRD people who primarily click on ads, hence contributing to Google's ad business and revenue. What Google later noticed is that the insights gained from these WEIRD users, whose behaviors and norms are more likely familiar to Google developers, don't necessarily generalize to other countries. It was only when Google tried to enter countries where a local search engine provider existed that it noticed this strategy might not always work out. The local companies simply had a significant advantage over Google because they intuitively knew market preferences and particularities and often had the trust and support of the local user population.

Traveling the Online World

Contrasting Google, Naver, and Yahoo! Japan gives us a neat illustration of how technology looks and feels differently across the world. It's like a kaleidoscope with its ever-changing, colorful patterns that you can look through and see different worlds. You can turn it, it may be blurry, and it may give you both positive and negative experiences. But it immediately shows you that there is more than one approach to a design. And, just like that, you don't have to travel far to experience cultural diversity today: Search for any given technology product and country combination (using the search engine and query of your choice), and in many cases you'll find a local version. I love this about the digital world. It's a much more carbon-friendly way to experience cultural diversity than physically traveling, and it can get you anywhere in the world within a few seconds (presuming a decent Internet connection).

If you are from Japan, for example, you may frequently use Mixi, a Japanese social networking site that launched in 2004 and has long dominated Japan's social media landscape. Or maybe you used to be regularly on Mixi until more and more of your friends joined Facebook and you switched because it made more sense with more of your friends on it. Economists call this a *network effect*, and it's the main reason why we have so few, but large, platforms in the current social media landscape. There needs to be a critical mass. When you finally gave in and created an account on Facebook, you were probably in for a surprise. Unlike Mixi, Facebook requires real names. It will immediately ask you to disclose a lot of private information. It will also encourage you to send out bulk invitations to your friends as a way to save time.[83] Real profile pictures are the norm and Facebook's tagging feature will make sure that soon everyone will see photos of you and know who you are.[4] People will also see your profile, or at least the parts that you have not painstakingly set to private, no matter whether they are friends with you or not.

All of this could be surprising to Japanese users who tend to be extremely private. Whereas American users love to convey their individuality—through photos, through clever news feed updates, and by sharing their relationship status, gender identity, political orientation, and eccentric music preferences—Japanese often prefer anonymity, are less assertive, and are less likely to share information about themselves online.[193] In fact, I'm pretty sure Facebook's business model of selling user data would collapse if they had only Japanese users: Their strategy of persuading users to tell the world about themselves simply clashes with Japanese cultural norms. What's interesting is that Facebook's and Mixi's designs mirror these cultural differences. They each work well in their local context, at least from their companies' perspective. And while strong forces like network effects may override preferences for one or the other, researchers have found that Japanese users simply tend to prefer Mixi.[4]

Researchers attribute these differences in online sharing to various cultural phenomena, among them a variation in individualism versus collectivism.* Have a look at this LabintheWild test (type the URL into any browser window) to see where you fall on the individualism/collectivism dimension and what this reveals about your online sharing behavior:

LabintheWild study #1: https://labinthewild.org/bookstudy1

What did your results say? If you found you are more collectivistic than individualistic, then you may have also seen that you find it more important to avoid privacy risks from oversharing information on social media than others. Researchers believe this may be motivated by safeguarding an ingroup, such as the reputation of your family and friends[56,280] (something that we will also see in Chapter 5 in the context of digital self-presentation). Likewise, if you scored high on individualism, you may be less worried about oversharing. I assume this to be the case because researchers have found differences between the privacy attitudes and practices of US and Chinese participants, with US participants commonly being less likely to restrict who can access their information on online social networks than the Chinese.[297] Participants from China, in contrast, have indicated a much higher desire than Americans (and participants from the Netherlands, Germany, and the UK) to have more options to address a specific group of people on online social networks.[279] This suggests that oversharing, to them, is of greater concern. Maybe you also fall somewhere in the middle or find completely different results for yourself; remember that neither these dimensions nor the relationships between the dimensions and online sharing behavior capture all of human diversity. There are often

---

*Another dimension, indulgence versus restraint, which Hofstede added in 2010,[126] has sometimes been found to explain privacy behaviors across countries and sometimes not.[30,157]

large deviations within countries, such as across socioeconomic backgrounds. For example, in Egypt and Saudi Arabia people from a low socioeconomic background have been found to be more at risk of, and more worried about, invasions of their privacy than those from a high socioeconomic background.[240] What this emphasizes is that people differ in their perception of privacy risks. Facebook's and Mixi's different strategies of persuading people to openly share their lives online will determine which of the social network platforms people will be more comfortable with.

The strategy of persuasion is also interesting to observe across e-commerce platforms in different countries. How do you make users feel more supported when they're shopping online? Or, asked differently, how do you get people to spend more of their money? Makuochi Nkwo and Rita Orji, both originally from Nigeria, have done extensive research on the topic of persuasion. In their prior work, they analyzed Konga and Jumia, two e-commerce giants in Nigeria and several other African countries that offer everything from computers and phones to toys, clothes, and household items.[199,200] Nkwo and Orji found that both of these platforms use personalization strategies that recommend products based on inferred user preferences. That is, Konga's and Jumia's strategy to get users to buy more is very similar to those of online shops around the world: Learn something about a user's age, gender, and other demographic variables and show them products that they may want to buy. But there is one aspect that may be even more powerful in persuading Nigerian users: the fact that they can connect with a chatbot. Known as JumiaBot on Jumia, the chatbot assists users in making choices by helping them decide which products to buy and how to buy them. Have a look at Figure 2.2 to see how these chatbots are readily available to guide people through the online shopping experience. Most importantly, the chatbots provide social support, something that researchers have found is one of the most important strategies for persuading people in collectivist countries like Nigeria.[199] Having someone to talk to offers a social bond between a platform and its users—a personal touch—which can be extremely important for collectivists.

A personal touch isn't important just on e-commerce platforms, but also in other settings in which motivation and persuasion play a role. Consider a study by Kiemute Oyibo and Julita Vassileva, who asked participants from Northern America and Nigeria to rank different persuasion strategies in fitness applications according to which would motivate them most.[211] Is someone more likely to go for a run if the application promises a reward? Or maybe competition? Setting one's goal individually? Or by cooperating with others? Oyibo and Vassileva found that Nigerians ranked cooperation and social learning highest—they felt motivated by working together with family and friends to achieve a collective goal and seeing the progress and achievements. North American participants, in contrast, preferred setting their own goals and working toward a reward to motivate themselves. The researchers' conclusion? Persuasion is much more interpersonal in Nigeria than it is in North America.

What I see in Konga and Jumia is that these platforms are designed to emphasize social support. It's not provided by friends and families, but nevertheless provides a personal connection and bond. This makes the platforms much more suitable for the Nigerian market than the non-African e-commerce platform Amazon, where finding anyone to talk to is quite a challenge. (And if you briefly stumbled over the mentioning of "non-African," you just evidenced the Western technology bias many people have.)

What do you think when you look at Jumia's and Konga's interface in Figure 2.2? Do the sites look anything like other e-commerce sites you may be used to? I have to admit, I was initially a bit taken aback. Those are some very busy sites, I thought to myself. There's lots going on: bright colors; attention-seeking, animated ads; lots of images; a patterned background—even the margin areas outside of the main content are covered in ads. Now have a look at Figure 2.3, showing the popular German e-commerce platform Otto.de and the Norwegian Komplett.no. They're not simple websites, as e-commerce platforms rarely are because they want you to immediately see something that you'd like to buy. But they are much more subdued than Konga and Jumia.

(a) Jumia.com.ng

(b) Konga.com

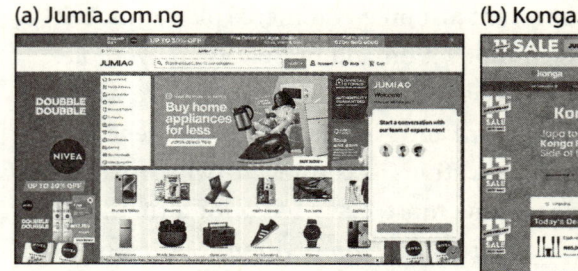

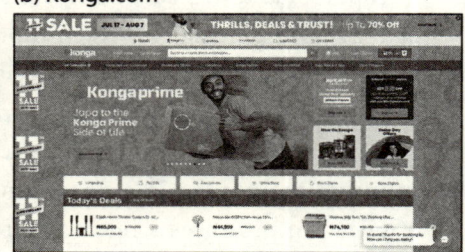

FIGURE 2.2. African e-commerce platforms Jumia and Konga.

I find the contrast in designs pretty interesting. It's just one example of how design choices can differ, but it shows how the Internet really is a kaleidoscope of technology. It also underlines that local technology products, including websites, can be popular for various reasons. Sometimes it is because they are the only ones to offer a specific service. Or they were the first to introduce a service, which can have a winner-takes-all effect. Maybe they successfully marketed their service and won the trust of people. Sometimes it's because the government championed local platforms and curbed those from foreign companies. And finally, many times it is simply because their product fits the local culture best.

Design in Context

The previous examples don't just tell us that there is some kind of elusive influence of culture on product design and success. They also tell us that context matters. A digital technology can only be successful (however you want to define that) if it fits the needs, desires, and norms of a specific user population. When I teach HCI, I try to make sure that students learn how to understand the goals and needs of their users within their particular context. Often, we use techniques from *Contextual Design,* an approach first proposed by Hugh Beyer and Karen Holtzblatt in 1999, which encourages designers to deeply immerse themselves in any aspect of the lives of the users they are designing for.[38] The American design and consulting company IDEO

(a) Otto.de

(b) Komplett.no

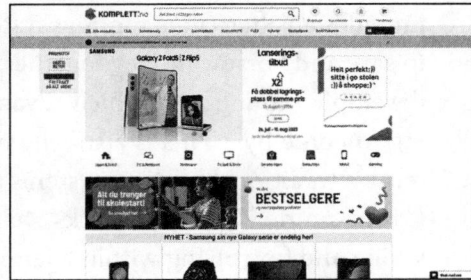

FIGURE 2.3. European e-commerce platforms Otto and Komplett.

describes this as developing empathy for one's users: "When companies allow a deep emotional understanding of people's needs to inspire them—and transform their work, their teams, and even their organization at large—they unlock the creative capacity for innovation." To achieve this deep emotional understanding, IDEO's designers have spent nights at eldercare facilities to experience what sleeping on rubber sheets feels like or gotten their chest hair waxed to understand the fear wound-care patients are experiencing when awaiting treatment.[32] Admittedly, these are hair-raising examples, but it's easy to see how such exposure could force designers to see the world through someone else's eyes. The problem we see may shift, and new solutions may become apparent. And in many cases, it would become obvious that some ideas are terrible. For example, if you try only once to work as a fisherman, wearing gloves to deal with lots of slimy fish and very cold water, you will very quickly notice that it's not a good idea to design an app for fishermen with incredibly small buttons on a mobile screen. Speech input may be more appropriate. And similarly, it would become quickly apparent why many technology ideas that seem useful in one country, such as a courteous robotaxi or a ridesharing app designed for solo travelers, may not be useful in others.

Across the world, we can now find technology products that are successful in only a few specific cultural contexts. They are invented to fulfill the needs in a specific kind of cultural context, and they generally won't see uptake in other cultures. For example, the Egyptian fintech app Money Fellows attributes its popularity to the fact that it provides an alternative funding system—supporting groups of people to save and borrow money together.[139] This works particularly well in developing countries in which advanced financial systems are challenging to come by. (This is also why mobile banking in Africa does not rely on traditional banking systems as Western apps such as Venmo or Cash App do.[20]) It also works well in collectivist societies in which saving and borrowing within a family or group of friends is common. But, as Money Fellows' founder and CEO Ahmed Wadi found, the demand for the app is virtually nonexistent in countries like Germany and the UK, where people can turn to financial institutions to borrow money and would often rather do so than ask friends and family for help.

For other examples we just have to look at the many technology products that serve different religions. Need a second wife? There's an app for that! It's called SecondWife, and it's a Muslim polygamy matchmaking service that is popular in Muslim communities in various countries across the world.[302] The Islamic cultural context is essential for its success—the app would undoubtedly fail in cultures in which polygamy is frowned upon. (As a result of these social norms, polygamy is illegal in many countries, though there are sometimes exceptions for subcultural groups.)

Design Localization

You might have noticed that the previous examples of technology were about products that were developed locally with a specific user population in mind. I don't think their design involved any particular hair-waxing experiences or similar efforts to understand users; instead, it's likely that the developers designed the products intuitively based on their own needs and cultural frames. Maybe they did some market research and user testing, but because their products were not designed to be global, they probably tested them on local populations, if at all. It's only when developers anticipate that their product will also be used in other cultural contexts that they sometimes think of making changes. They may also see the product spread across countries and cultures and, in hindsight, realize that it isn't fully useful for others, which could encourage them to adapt some aspects of the technology. Making such changes is called "localizing" a product, and it is most often used to refer to offering additional languages and changing a few things, like date and time formats, that often don't translate well across countries. Larger, globally operating companies sometimes go beyond these basic aspects of localization and offer products that take into account more hidden dimensions of culture. Google's attempts to make its search engine web page more playful and add functionality for the South Korean market is an example of this. It is sometimes referred to as "design localization" because it is about adapting various aspects of the user interaction design. And while it can be time-consuming and expensive, many global companies know that truly localizing their products beyond just adapting the language can make or break their brand.

Let's stick with e-commerce for a moment, since we were talking about it earlier. Last I checked, Amazon offered 22 different country versions of its website—an indication of its efforts to maintain or gain market share around the globe (though 22 website versions sounds somewhat pathetic if you compare it to Coca-Cola, which maintains 115 versions for different countries, sometimes with several language versions per country).[14,59] Have a look at Amazon's US, German, and Saudi Arabian websites in Figure 2.4 (or you can also go to the live versions using its country selector) and you'll probably immediately notice several striking similarities and differences. For one, it's easy to see that all three versions are Amazon sites, an important goal for brand recognition. They all have the iconic white and orange logo, black menu bar, gray background. A prominent search bar at the top of each of the sites and a list of products in the main content region immediately give away its purpose as an e-commerce site. But let me point out some of the differences. Of course, there is the language adaptation—the cornerstone of any localization effort. On top of that, Amazon also adapted the content (check out the advertisement of Hispanic goods and pickleball equipment in the US version—both things people in Germany and

(a) Amazon.com

(b) Amazon.de

(c) Amazon.sa

FIGURE 2.4. E-commerce platform Amazon and its US, German, and Saudi Arabian versions.

Saudi Arabia are less likely to relate to). In the Saudi Arabia version, Amazon changed the script and reading direction from right to left: The logo is now on the top right, while the shopping cart is on the left, in line with how Arabic readers would view the page and expect it to be arranged. (Don't underestimate these adaptations: Obviously it's important to be able to read a website in a language that one understands and reads well, but researchers also think that the fact that much of the Internet is dominated by the English language and the Latin alphabet with a left-to-right reading direction has contributed to the relatively late uptake of the Internet among Arab countries.[301]) Beyond language, script, and reading direction, Amazon's designers also played with the visual complexity of the design. In the German version, they reduced the clutter common to most e-commerce websites by narrowing the main content area (thus leaving more screen space to a simple gray background) and by removing the sub-menu bar such that there is only a narrow black menu at the top. This is much more in line with other German e-commerce shops, such as the Otto.de website we looked at earlier, and it's also consistent with Germans' tendency to focus on utility and a streamlined design.[61] Also, where's the icon for the shopping cart? And where's the in-your-face-suggestion to "sign in for the best experience"? For the German version of its site, Amazon must have thought these persuasion strategies are too aggressive for new German customers, who tend to be cautious and restrained online shoppers.

Research has provided much evidence for the importance of localizing products beyond just adapting the language. We'll discuss this more in Chapter 6, where you'll learn that design localization can improve user satisfaction and work efficiency, because it can make products more intuitive to use. In addition, websites that are culturally adapted can also increase people's trust in them, which, unsurprisingly, is highly important if people are supposed to rely on the website's information and buy something.[94] For example, in a study with undergraduate students from Italy, India, the Netherlands, Switzerland, and Spain, participants indicated that they would be more likely to purchase a product from highly adapted e-commerce websites (US company websites that had specifically designed country versions) than from websites that showed minimal adaptation.[252] But their ratings also suggested that even better than localized websites are *local* websites provided by companies from the same country as the user—an indication that design localization can only go so far.

So what's your guess in terms of today's Internet landscape? Do you think the most popular websites are local websites or those that are run by globally operating companies? I can tell you that, as of five years ago, people spent a big chunk of their time on local websites. In fact, when my lab and I analyzed the 2,000 most popular websites across 44 countries, we found that only a small percentage of them (around 80 or 90 per country, or roughly 4%) were global websites, like Google and YouTube.[202] But this also differs by country. For example, of the 100 most popular websites in a country, the Netherlands has the highest percentage, 56%, of global websites. In contrast, only 15% of the top 100 websites in Russia and 20% of those in China can be considered global sites. You can see how this would dramatically influence how much "global design exposure" people get—or maybe I should call it "US design exposure," since many of the global websites were designed by US companies and didn't substantially differ in their visual designs across countries. Indeed, our study confirmed what we saw earlier with the Amazon website versions: Global companies make some efforts to localize, but ultimately they rely on relatively homogeneous website designs. Local website designs, in contrast, often look quite different from the website designs of globally operating companies.

These local websites are what we should view through our kaleidoscope of cultures in the online world. Look at them instead of global, localized website versions to understand how beautifully diverse technology designs can be across countries and cultures.

Design localization is important not just for the visual aspects of a product. Several years ago, my former PhD student Judith Yaaqoubi and I wrote about the various localization efforts for Cortana, Microsoft's former personal productivity assistant.[318] It was recently replaced by a new and shiny AI, but back in 2014 and for a number of years after that, Cortana was quite a hit. It could set reminders, track packages, identify travel plans from a calendar app, and complete a few other useful tasks that one could ask it to do via basic natural voice commands. It also answered questions about the weather, the newest sports results, or similar things by accessing the Bing search engine. Microsoft quickly offered it in several other languages and with different accents, such as Indian English, Mexican Spanish, Brazilian Portuguese, and Canadian French. The company even set up a Cortana Personality Design Team to give it a "personality"—after all, Cortana was actually talking with people who tend to humanize AI. But then came the difficult questions: What kind of personality should Cortana have? How should it present itself to its users? The Personality Design Team agreed that Cortana's personality needed to be adjusted for different markets, each representing specific cultural expectations. US Cortana, they decided, should be professional, helpful, and good at small talk. When starting a conversation, it should frequently talk about "her" history, family, or taste in music—all ways to become more personable and build trust, the Cortana Personality Design Team reasoned, in line with Americans' high individualism. (And of course the default gender was initially female, but that's a topic for a different book.) The Indian team members felt that Cortana should be more collectivistic, by talking less about itself and more about local foods, rituals, and celebrations. And the French team members felt that Cortana should be more formal when speaking with French users, including addressing users with the formal "you" ("vous").

What was interesting to see was that the Microsoft team tried to incorporate Hofstede's cultural dimensions into what Cortana's personality should be like in different countries. This gave them a way to systematically think about cultural differences and sort of acted as a starting point to brainstorm about Cortana's design localization. But the team also heavily relied on the personal experiences and viewpoints of team members from a certain country to know how to design for that market. They had to do this because the cultural dimensions don't give specific design advice. You can use them to refine or justify an assumption, but the specific design decisions are still going to be influenced by people's intuition.

I think what the Cortana team experienced here is one of the key problems with designing technology for other countries and cultures: It is really difficult to know what subgroups of people may like. This goes back to the WEIRD problem in science I told you about in the introduction, and to the issue that humans are simply so diverse that even within a fairly homogeneous group of people, one size will not fit all. How someone working for an international corporation like Microsoft might think Cortana should act could be completely wrong for others, even if they are from the same country.

Unfortunately, I'm not aware of any published studies that have evaluated whether Cortana's multiple personalities—and Amazon's many website versions—were ultimately successful in different countries, so I cannot tell you whether these design localization efforts were actually worthwhile. I would be surprised if the two companies hadn't done quite a few studies ("A/B tests," as companies call it when they give a subset of users version A of a product, another subset version B, and then compare usage patterns or other metrics). Alas, companies have no incentive to share such results with the public and their competitors, so we may never know. But even without knowing whether the localized version improved some kind of metric, I think it's undisputed that localizing the way technology communicates with us is important— not only for ensuring equitable access but also for avoiding failures in additional markets. As a constant reminder of this risk, the Internet is full of localization blunders, many of which involve mistranslated slogans: my favorite ones are the Swedish company Electrolux's trying to sell its vacuum cleaners with the slogan "Nothing sucks like an Electrolux" and Pepsi's "Come alive! You're in the Pepsi Generation!" which translated into Mandarin means "Pepsi brings your ancestors back from the dead!" Some of these fiascos are now regularly taught in business school courses. For example, business school students often learn about Procter & Gamble's marketing of disposable diapers being delivered by a stork, and that people in Japan had no idea why a stork, and not a giant peach as in Japanese folklore, would deliver diapers.

While this happened before the Internet, the implications are still the same: Words, folktales, the meanings of colors, symbols, and images—all of these can easily be lost in translation. Hearing about, seeing, or even using technology designed in other countries and cultures can be a sudden eye-opener that the design choices we have long assumed to be the norm are not the norm elsewhere in the world.

3

Perceiving Information
in Different Cultures

In 2000, neuroscientist Eleanor Maguire and her colleagues at University College London published an influential paper that reported on their study of taxi drivers in London.[173] If you've ever tried to navigate your way around London, you'll know that the city consists of endless winding streets whose names do not follow any particular pattern. Unsurprisingly, becoming a taxi driver in London requires extensive training to learn how to get to thousands of places in the city. Aspiring taxi drivers often study two to four years to memorize various routes before they are licensed to operate—until they have finally mastered "The Knowledge," as Londoners refer to it. Last I checked, passing "The Knowledge" exam was still a requirement, even in times of Google Maps and the like. Do the brains of these future taxi drivers look different than the brains of people who did not rigorously train for spatial navigation?

In their study, the researchers compared the brains of 16 male licensed London taxi drivers to those of 50 male control subjects using structural magnetic resonance imaging (MRI) scans. This essentially let them analyze the cortical gray and white matter volume that the brain is composed of. The results were fascinating. Analyzing the scans, Maguire found that, compared to the control subjects, the taxi drivers' brains had a much larger posterior hippocampus (the region of the brain that is thought to be involved in spatial memory and navigation).

On top of that, the longer they had been a London taxi driver, the larger this part of their brain was. (Their results suggested that the overall brain was not necessarily "bigger," but that the gray matter in the hippocampus was redistributed based on the experiences.)

What makes Maguire's study so interesting is that it shows how our brain is malleable over the course of our lives, even if we're just learning a new skill. Study London's roads for two years, and you, too, will increase the volume of your posterior hippocampus, her results suggest. But there is something even more exciting about it that I want to point out. Think, for a moment, of London taxi drivers as a cultural group within the UK. A subculture. The way that studying for "The Knowledge" is reflected in neuro-anatomical changes in their brains suggests that cultural practices—experiences—can shape our brains. Just like the differences between the brains of London taxi drivers and the control group, we could expect to see differences between the brains of people who grew up in distinct cultures.

There is no evidence that human behaviors, or our brains for that matter, are different when we are born. But our learning, our cultural experiences, shape how we think and behave over the course of our lives. According to psychologist and neuroscientist Lisa Feldman Barrett, even our emotions are formed by these experiences—rather than being automatic, pre-programmed responses of our brains, as has long been assumed (and is still being debated).[29] Anything we experience during our formative years, roughly before we turn eight, likely has the largest impact on our brains. So wherever you've spent those years, and whatever the cultural norms you were exposed to through your interaction with your family and society, those will have tremendously shaped your brain.

Foreground or Context

This idea is especially fascinating because researchers studying the brain and cognitive processes have traditionally not looked at cross-cultural differences. Even today, the equipment needed to use neuroscience methods is more commonly available in rich countries, such as Western industrialized countries. Take this as another area of science where most of our knowledge stems from studying WEIRD participants.[55]

Luckily, some researchers suspected early on that cognitive processes, such as perception, attention, and memory, may be shaped by culture—and their assumptions were confirmed across many behavioral studies. One of the earliest of these studies was by cross-cultural psychologists Takahiko Masuda and Richard Nisbett. They asked Japanese and American participants to watch short video clips showing underwater scenes and report on what they saw.[182] When reporting back, Japanese participants mentioned 60% more context information, such as objects that did not move or that were in the background, than Americans. In contrast, Americans generally described what they saw starting with salient objects, such as the fish in the foreground. Does this also influence what someone actually perceives and remembers? To find out, the researchers showed participants objects that had previously appeared in the short videos, either with no background, the original background from the prior video, or a new background. No matter what the background, American participants in the study were able to equally detect whether they had previously seen the objects. But Japanese participants had trouble doing so when the objects were placed in front of new backgrounds. Because they had focused more on background and nonmoving objects in the videos, they were later better at identifying objects placed in front of an already familiar background.

Nisbett and his collaborators defined these differences as analytic versus holistic.[198] According to their definition, people who focus on objects and the specific characteristics of these objects tend to be analytic thinkers. For them, objects are mostly seen as independent from their contexts. Holistic thinkers instead relate objects and the context to each other, which is why the Japanese participants in Masuda and Nisbett's study perceived the entire image as one and had trouble recognizing parts that were presented with a new background.

Of course, the question is whether the Japanese really did focus more on the background than Americans, or whether they just reported on the scenes in a different way. To find out, Nisbett's lab followed up on their previous work with an eye-tracking study, this time comparing American and Chinese graduate students at the University of Michigan.[57] The team asked participants to view 36 pictures, each with a single foreground object (an animal, boat, car, or plane) on a realistic background. After seeing a picture for 3 seconds, participants rated it on how much they liked it. (This was primarily done to get participants to engage with each picture.) Sure enough, Americans focused their eyes more on the foreground objects than Chinese participants. They looked at them more quickly at the beginning as if their attention was automatically drawn to those objects. They also looked at them for longer periods of time than the Chinese. In contrast, the Chinese more often quickly moved their eyes between the foreground object and the background, seemingly binding the elements together. The authors concluded that the two groups "allocated attentional resources differently as they viewed the scenes." In other words, there is a difference in how people observe information, and not just in how they report on it.

And this difference doesn't just exist when people view pictures. A few years after this finding, Ying Dong and Kun-Pyo Lee from KAIST in South Korea wondered whether the differences in viewing patterns across cultures would show up when people look at websites.[68] The researchers asked Americans, Koreans, and Chinese to freely look at an imitation of the Yahoo! website in their native language. Participants had 30 seconds to do so, which should give a good idea of how they initially approached a website. I found the results especially intriguing because they revealed tremendous differences in where on a website people place most attention. Just as when viewing pictures, Americans tended to view the different parts of the page in sequential order, seemingly treating each part of the page independently. Contrary to this analytic approach, most of the Korean and Chinese participants viewed the website in a nonlinear way, scanning back and forth between parts of the page. Websites are of course very different from the pictures used in the previous study. For example, there is no clear foreground object in most cases. But the eye movements by the Chinese and Koreans suggested that they were visually creating connections between different website areas.

A few years ago, my lab and I, together with our Japanese collaborator Naomi Yamashita, conducted an online study to find out whether this difference in attention patterns could affect how people find information on a website.[34] If East Asians approach websites more holistically, scanning a page for relationships between objects, how does that impact how quickly they find information? And given that Westerners seem to focus more on foreground information, does that mean they are slower at finding information on the periphery than East Asians? Our study asked participants from Japan and from the US to complete different scenarios, such as opening a bank account for which they first needed to register. After reading through the scenario description, they were shown a website (in Japanese or English) on which they had to find the button for registering. One of our assumptions was that Japanese participants would be faster at finding information on visually complex websites—after all, many Japanese websites are more complex than those in the US.

What we actually found was the opposite. While both participant groups were equally good at completing the tasks, Japanese participants took quite a bit longer to find information than the Americans. In fact, they took three times as long on visually complex websites than American participants, no matter whether the information they had to find was in the middle of the screen (what we had defined as "foreground") or on the periphery. What was going on here? Why did the Japanese take so much longer? We considered the script, since Japanese characters are slightly slower to read than the English characters. (Japanese readers read an average of 193 words per minute, whereas the average is 228 words per minute for English readers.[278]) Yet, we found that our Japanese participants were also slower finding information on websites with very little text, so this was unlikely the cause for the delay. It could be that, by giving participants the task of finding only one piece of information and ignoring all others, our study was inadvertently favoring Americans. But then there is another potential explanation: Recall that some of the previous work had found that Japanese were making very fast, non-targeted eye movements when looking at scenes. Researchers have interpreted this to mean that they were establishing the relationships between objects. Americans in these prior studies fixated sooner on specific objects and were later more accurate in recognizing them. Our interpretation of the combined results is that, rather than immediately seeking out information as fast as they could as the US participants did, the Japanese participants seemed to be taking the extra step of holistically scanning the website. Only after making sense of the previously unknown website did they engage in the primary search task. We can't be certain about this, given that our online study did not allow for the use of eye trackers. But what we know right now is that Japanese participants might familiarize themselves with new websites more holistically, which simply takes more time.

What causes these differences in attention? This is where our scientific knowledge gets even more wobbly. According to cross-cultural psychologists, one possible explanation is that there are differences in social structure and practices as a result of ancient civilizations and their philosophies and moral thoughts. (Two different civilizations that are often cited here are those of the ancient Greeks, with their sense of personal agency, and Confucianism's Chinese, who emphasized collective agency and social obligation, and also heavily influenced the culture in Japan. If you are interested in reading more about their fascinating philosophies, I encourage you to read Nisbett's book *The Geography of Thought*,[197] or one of his articles, such as the one on "culture and systems of thought."[198]) Perhaps as a result of these different philosophies, people's self-construals differ across cultures: Westerners perceive themselves as more independent, while Confucianism has promoted a more interdependent, relationship-focused self-construal.[178] And this difference is even noticeable in how people talk to their children (for example, Japanese parents have been found to describe relationships between objects more often than US parents, who tend to focus on teaching nouns first), and how we later categorize words: While Americans more often combine words based on a taxonomic understanding of the world (given the words monkey, banana, and panda, they would likely combine monkey and panda as the two animals in the list), Chinese were found to combine them based on their relationships ("monkeys eat bananas").[134]

Scientists are of course often uncomfortable claiming causality. There is still so much we don't know about humans and their complicated brains. But let me summarize our current understanding of things: It is very likely that Americans learn to mostly focus on the fish, on other primary objects, and on specific website functionalities because they have been taught to see themselves and other things in this world (the fish in this example) as independent. East Asian cultures instead teach children to pay attention to the relationships between people, objects, and the contexts they interact with.

But there's also some indication that our exposure to specific environments can change this, at least temporarily. In 2006, Miyamoto, Nisbett, and Masuda found that common city scenes in Japan are much more cluttered than the equivalent scenes in the US.[186] To determine this, they took pictures of the scenes around schools, post offices, and hotels in comparable cities of different sizes in the two countries. The pictures taken in Tokyo, for example, were much more complex than those taken in New York City. The authors then used a procedure that psychologists refer to as "priming." Participants were shown different primes—either pictures taken in the US or ones taken in Japan—before participating in a change blindness task, which asked them to view two similar images, presented in succession, and identify any changes. Imagine this being similar to a "spot the difference" picture puzzle for kids.

If it is really true that attentional patterns are trained rather than our being born with a certain way of looking at scenes, the authors reasoned, then looking at the more complex Japanese pictures should make both American and Japanese participants slightly more holistic than when they are looking at simpler US scenes. Their assumption was confirmed: When participants were shown the more cluttered city scenes in Japanese pictures, rather than pictures taken of US cities, they later detected more contextual changes in the change blindness task. And this was true for both American and Japanese participants.

Think about this for a moment. The finding by Miyamoto and his colleagues basically suggests that if you spend some time in an environment that contains more objects—let's say, in a cluttered room or on a busy street with neon advertisements—you will become better at detecting changes in the context of an image. You will essentially become less change blind! And maybe—just speculating here—you will become more distracted by those notifications that sometimes pop up on the periphery of our screens. Honestly, I cannot imagine the impact to be profound, but this is nevertheless a fascinating topic to think about. Our brains are not only influenced by our exposure to cultural norms and values, but they can also be temporarily primed such that we focus our attention differently.

Seeing Culture in the Brain

If you're as fascinated as I am about our malleable brains, read on. Let me tell you about a research area that I have followed closely ever since I met cross-cultural psychologist Shinobu Kitayama at the University of Michigan. A few years before I arrived in Michigan for my first faculty job, Kitayama had embarked on a new research direction in cultural neuroscience.[147] The premise of this relatively young field of research was that scientists could finally explore whether differences in behaviors, attention, and memory could also be seen in the brain. As Kitayama once explained to me over coffee in a small cafe on Ann Arbor's main street, this was especially important because it was still an open question whether brain processes are universal. After this field started, it soon became obvious that our malleable brains change based on cultural experiences.

In 2006, psychologist Angela Gutchess and her colleagues were some of the first to contribute to this exciting new area.[108] Their study was very similar to the one I described earlier—participants saw pictures that showed only a foreground object on a white background, only the background, or both combined. Each time, after seeing a picture for a few seconds, they were asked to rate how pleasant they found it. The key difference from previous studies was that participants were

performing this task in a full-body MRI scanner. For each of her participants, the scanner gave Gutchess images showing slices of their brain,
which she could then analyze to see which areas of the brain were most
active. The results were as expected but nevertheless exciting: American participants activated more brain regions that are involved in the
semantic processing of objects—suggesting an increased attention to
objects—than did East Asian participants (who were originally from
China and Hong Kong). Even at a relatively young age—participants
were between 18 and 29 years old—and even though all participants
currently lived in the US, their cultural experiences in their home
countries had shaped their neural activity.

There's one more exciting piece of evidence that I would like to show
you. But first, I want to encourage you to try out a test on LabintheWild,
the framed-line test. In this test, you will see several boxes—frames—
drawn on the screen, one at a time. At first, you will see a frame with
a line drawn in it. On a subsequent screen, you will then see an empty
frame of a different size, in which you should draw the line. Pay attention to the instructions: In one part of the study, you need to replicate
the *exact* length of the line, so you will need to ignore the size of the
frame. In the other part of the study, you need to draw the line so that
it is the same length *relative* to the size of the frame. Whichever version
you're asked to do first is randomized across participants, which means
you may be asked to first draw the relative length and later the exact
length or vice versa.

LabintheWild study #2: https://labinthewild.org/bookstudy2

How did it go? Did you find it easier to draw the exact length of the
line or the relative length? Did the test tell you that you are more Eastern
or more Western? To give you a bit of background, this study was first
invented by Shinobu Kitayama when he was still at Kyoto University in
Japan, together with several colleagues in the US.[145] The reason they
invented it was to find out whether East Asians are better at incorporating contextual information (which would suggest they are more holistic) and whether North Americans are better at ignoring contextual

information (which would make them more analytic). In contrast to previous studies, they wanted to test this in a single study and use an abstract concept, a simple frame and line. Removing images of scenes and objects from their study and instead resorting to an abstract frame and line had the advantage that they could potentially rule out that any observed differences are due to familiarity with the scenes and objects, rather than due to cultural differences in perception. What they found was quite eye-opening: US participants (students from the University of Chicago) were better at ignoring the frame when replicating the exact length of the line than Japanese participants (students from Kyoto University). But Japanese participants were better at incorporating the size of the frame and replicating the line relative to the size of the frame.

If the LabintheWild study told you that you are more Western, it means that you were better at replicating the exact length of the line and ignoring the frame. This is your culturally preferred task. If your results suggested you are more Eastern, it would mean the opposite: You must have done better when drawing the line relative to the size of the frame and may be more prone to connect foreground and background objects (the line and frame).

Now, if I had asked you to do the same study in an fMRI scanner, I could have told you what your performance in this study looked like in your brain. Unfortunately, very few people have fMRI scanners lying around as they are extremely costly. But in an article published in *Psychological Science*, a research team led by Trey Hedden at Stanford and MIT showed what would happen if one did.[120] In their study, they scanned the brains of 10 Americans of Western European ancestry and 10 East Asians while they were completing an adapted version of the framed-line task. (Instead of drawing the lines into an empty frame, participants were asked to judge whether a frame with a line matched one that was previously presented to them.) When the Americans judged whether the line was the same relative to the size of the frame—the culturally non-preferred task—the research team saw that they were activating several areas of their brains more than when judging in the culturally preferred task, in which they could ignore the size of the frame. For East Asians, it was the exact opposite. Upon a closer look, the research team found that it was the pre-frontal and parietal areas of the brain that were firing whenever participants would work on the culturally non-preferred task. They are the parts of the brain that really get going when we are working on very demanding tasks. So the exciting discovery was that Hedden and his team could see the extra effort it takes to carry out the culturally non-preferred task in the brain—a sign that there is a neural cost to working on tasks that are less "culturally natural" for us.

Organizing Information

There is other evidence that we learn to perceive and process information differently and that there is a neural cost if we have to adapt to ways unnatural for us. Take how we organize information as an example. Information is often arranged to indicate a process—an implicit timeline that tells us what to read or do first and what next.

There are a few assumptions that we can make about how people mentally represent such timelines. One is that we conceptualize time using spatial representations. This is the case in cultural artifacts that tell us the time—think clocks, timelines, calendars, or even hourglasses where time passes as the sand vertically filters through. But we also use space to think of time in our minds.[41] Time is an abstract concept, of course, yet we create spatial metaphors for it based on our experience.

In my mind, temporal events are always arranged in a timeline. As a reader of the Latin left-to-right alphabet, I was taught that earlier events should be on the left side, while later events are toward the right. Any timeline I'd ever seen in books, in magazines, or on the Internet was designed that way. In my world, an interface guiding me through a process, such as that of going through the check-out funnel on an online shopping website, usually describes stages that happen first on the left and later ones further on the right. In addition, something I only recently realized is that I also tend to think of time as *relative* to my own position. An event that has already happened is to the left of me, and something that is going to happen, or that I should do in the future, is to my right. Maybe you use the same egocentric spatial representation when you think of time. But what I've learned is that it's not a universal phenomenon.

In 2010, cognitive scientists Orly Fuhrman and Lera Boroditsky published an article showing that Hebrew speakers arranged temporal information the same way as they read, right to left, while English speakers working on the same task arranged information left to right.[88] The reading direction seems to affect how people spatially represent time. In additional experiments, Fuhrman and Boroditsky then showed that if presented with temporal information, participants took longer to decide which event was earlier if the arrangement was incongruent with their reading direction (so, left to right for Hebrew speakers and right to left for English speakers) rather than congruent. Just as in the frame-line study by Hedden that I described earlier, they found that it is more difficult for people to interact with information in ways that are not culturally natural for them. Because of these results, I suspect that there is again a neural cost to people encountering user interfaces and visualizations that do not follow their reading direction. It likely takes them more time and effort to convert the temporal information to their normal reading direction in their head—another reason why design localization is so important.

Reading direction is not the only aspect that determines our spatial reasoning. The linguist Alice Gaby actually found that speakers of Pormpuraawan languages like Kuuk Thaayorre, a language spoken in the remote Aboriginal community of Pormpuraaw in Northern Australia, do not have any terms that describe space as relative to themselves, such as "left," "right," "in front," or "behind." For speakers of English, Dutch, German, Japanese, and many other languages, this is very difficult to imagine, because they think of the position of things around them as relative to themselves. When we turn, a tree may no longer be in front of us, but maybe it's now to the left. When we describe the position of the tree to others, we might say: "Look at the tree on my left." In contrast, speakers of Kuuk Thaayorre describe objects in space using an absolute frame of reference: "The tree is south south-west of me." They refer to the four cardinal directions, as well as to the north and south banks of a nearby river to describe both the position of objects to each other and geographical locations.[91] In her doctoral dissertation, "A grammar of Kuuk Thaayorre,"[92] Gaby illustrates just how different, and potentially consequential, this way of referencing objects could be, with the example, "The glass of poison is to the east of the glass of wine." Not particularly intuitive for those of us who are used to a relative frame of reference! Needless to say, I would have to think about whether I really wanted that glass of wine. The fact that so many mnemonics exist for remembering the cardinal directions in various languages suggests that others may hesitate, too—for many, the absolute frame of reference is too rarely used to become intuitive.

Given that speakers of Kuuk Thaayorre need to stay oriented at all times when talking to each other, Boroditsky and Gaby wondered whether this may influence how they think about time. They asked Pormpuraawan participants who were fluent in English as a second language to arrange different sets of cards in a timeline.[42] Each set of cards showed a temporal progression; for example, a series of images, each printed on a different card, showed a person from a young to an old age. To test the effect of their absolute frame of reference, each participant was arranging the cards while facing two different cardinal directions. (Whenever this was done on the same day, the experimenters asked participants to change direction to get a better camera position, so that participants wouldn't know that their spatial orientation was the focus of the study.) For comparison, Boroditsky and Gaby repeated the experiment with American students, all of whom laid out the cards from left to right independently of the cardinal direction they were facing. But most of the Pormpuraawan participants organized time from east to west. For example, when facing east, they would organize time as coming toward them; when facing north, time would be arranged from right to left; and so on. Moreover, the Pormpuraawans were also incredibly good at knowing where north, south, east, and west are. In fact, Boroditsky and Gaby found that they were only ever 10° off compared to the correct cardinal direction. American participants were all over, clearly not very attuned to thinking this way. I wouldn't be either, but at least I have the excuse of having spent much of my life in locations that rarely see the sun.

Boroditsky and Gaby's work showed that Pormpuraawans refer to cardinal directions when they describe locations and objects in space, and that they rely on cardinal directions when arranging temporal events. In a later article, appropriately titled "The Thaayorre think of time like they talk of space," Gaby set out to disentangle whether there is a causal relationship between describing space with cardinal directions and mentally representing time.[91] The way Pormpuraawans use an absolute representation of time is not necessarily a product of language, she wrote. It could also be due to something about their environment or their social and cultural contexts. To shed some light on this, Gaby repeated the previous study, this time comparing two small groups of Pormpuraawans: ethnic Thaayorre who are bilingual in Kuuk Thaayorre and English and those who are monolingual English speakers. This way she was able to hold constant the potential influence of their environment and their cultural context. And indeed, the responses of the monolingual English-speaking Thaayorre were indistinguishable from those of the American group in Boroditsky and Gaby's previous study. The monolinguals consistently arranged events from left to right. The responses of the bilingual Kuuk Thaayorre/English speakers instead tended to show the east-to-west pattern found in the previous study. This doesn't necessarily mean that there is a causal relationship between language and thought, as Gaby discusses in her paper. But her study does show that English-speaking ethnic Thaayorre represent time differently from those who are bilingual in English and Kuuk Thaayorre.

It turns out, there are several other languages that primarily use an absolute frame of reference, such as Tzeltal (Mexico), Guugu Yimithirr (Australia), Hai//om (Namibia), Longgu (Solomons), Balinese (Indonesia), and Belhare (Nepal).[174] So, really, using an absolute frame of reference is distributed across various language families around the globe. There are also languages that use an object-centered frame ("The glass of poison is left of the wine," where "left of" is interpreted from the perspective of the wine glass), such as in the language of Mopan (Belize) and Totonac (Mexico). And there are many languages that use all three frames of reference fairly equally, such as Ewe (Ghana), Kilivila (Papua New Guinea), and Tiriyó (Brazil). In short, there is immense diversity across the world in what spatial frame of reference people are most used to using. Contrast that with conventional user interface designs: No matter which way we're facing, they will present temporal information (from calendar items to a step-by-step guide) in the same way. No matter which way we hold our cell phones, it's still the same.

There are many other studies that have explored the relationship between language and its effects on our cognition. The assumption researchers often make is that language shapes the way we think—a theory that is called *linguistic relativity*. Proponents of this theory believe that language can also direct our attention to different colors, which could mean that some of us are better at categorizing and judging the similarity of some colors. Russian speakers, for example, were found to be faster when distinguishing between two blue tones that the Russian language has different words for than blue tones for which there is only one word.[309] In addition, a PNAS paper by researchers at the Max Planck Institute showed that speakers of Dutch and ≠Akhoe Hai//om, a hunter-gatherer community in Namibia, were better at solving a spatial relational learning task when using their preferred frame of reference (relative and absolute, respectively).[118] The authors concluded that human cognition is neither a "blank slate," nor uniform across humanity, as people sometimes assume. Instead, they argued, it is shaped by language and culture.

Now, as with most scientific theories, there is some controversy. There are many researchers who oppose the idea that language influences cognition[82,217]—their opinion is that speakers of different languages do not fundamentally differ in their perceptual processing. I'm not an expert on this topic, so I tend to lean back and watch the argument unfold. The Wikipedia page on linguistic relativity (and the corresponding Wiki Talk page) gives a fascinating overview of the different dueling opinions.

Inspiration for New User Interface
and Interaction Paradigms

I'd say, no matter what language has to do with it, there is a lot of evidence that differences in people's spatial reasoning and how they think of time across cultures are learned over time as a product of their lived experiences. But, of course, today's technology designs ignore the fact that some people think of time and space in an absolute way rather than relative to their own position or other objects. And who can blame them? We simply don't know very much about human variance, and we know even less about what this means for technology design.

So what I think we ought to do is take the different pieces of knowledge we have to date and build on them. We now have good evidence that people focus their attention on different things, some of them taking in images, scenes, and user interfaces more holistically than others. Some people are more susceptible to noticing changes in contextual regions than others, though, as we learned, people can be primed to be less change blind if exposed to visually complex environments. People also categorize things differently, either in a more taxonomic way or by focusing more on relationships and functions. And they use different frames of reference for arranging temporal information: Some do it based on writing direction in an egocentric frame of reference while others will order things according to cardinal directions. We've also seen that it takes people extra cognitive effort to perform a culturally non-preferred task and that this neural cost is measurable in increased brain activity. In other words, when we use digital technology that is not designed for us, our brain has to work harder to resolve the incongruencies and override what is more culturally normal for us. Today's technology is clearly biased toward Western cognition.

All of this research provides immense opportunity for rethinking how we design technology today, especially technology for which patterns of design and use have yet to be fully established. What, for example, if we took the insights I described above and applied them to the idea of spatial computing? How would we design the user interface and interaction if information didn't automatically flow left to right? I propose we leverage the insights into the diversity of human cognition to boost our creativity and question some of the assumptions we have made in prior user interface designs. Maybe interface components should actually rearrange themselves based on the direction we are facing, whether we're staring at our mobile phone or wearing a virtual or augmented reality headset. Maybe workflows that currently walk us through different steps should not automatically assume a left-to-right orientation. Maybe our files and menus should be sorted and arranged differently. And maybe the placement of general information and notifications should be rethought to optimize it for people's different attention patterns. I think there are opportunities to reimagine conventional graphical user interfaces, those that constrain us to viewing information in 2D. But I think the opportunities are even greater for newer virtual and augmented reality technologies where we could take advantage of space to support human cognition. People's perceptions vary and are continuously influenced by their lived experiences. It is time to use this to our advantage.

4

Technology Use around the World

In 2005, I went for a six-month-long stay in Rwanda to work on a software development and research project run as a collaboration between my local German university and the Rwandan agricultural ministry. With a laptop as heavy as an iron pan (it was my first laptop back then), I made daily trips to the only Internet cafe in downtown Butare, a small university town in the south. I still vividly remember climbing up the steep dirt road toward the main street, past little wooden corner stores that were just big enough to sell a few soda bottles, then past the local convenience store, a restaurant, and a seamstress's shop, in front of which I regularly saw women ironing clothes with heavy cast metal irons, who would more or less happily greet me with "Mzungu" ("white person"). If electricity was on at that time (power outages were common and would often last for hours), I would see the glowing monitors in the Internet cafe from the distance, an open door inviting people to come in and use the Internet for a fraction of a dollar per hour. Once settled into one of the cafe's chairs, with an Ethernet cable plugged into my laptop, I would frantically rush to send off pre-written emails, search for any information I urgently needed for my work, and check the news. The last thing I wanted to happen was not to have checked off all the "Internet to-dos" on my list before the power turned off again.

Of course, it did happen several times that the power went out in the middle of my sending those emails. But when I looked around, I was the only one who had exasperatedly thrown her hands up and experienced a mild panic attack about how the lack of an Internet connection would impact my work over the next few days. Everyone else . . . just leaned back in their chairs and started chatting, their laughter telling me that a power outage was nothing worth fretting over.

How could they be so relaxed about being interrupted in their tasks, especially not knowing when they may be able to finish them? In addition to being frustrated, I was simply stunned by their relaxed attitude. Only many years later did I find out that there is a psychological theory describing what I had witnessed. Edward Hall, a greatly influential and well-known anthropologist, had written a whole book, *The Silent Language*, about the rules of social time.[112] In it, he described how the perception of time is invariably intertwined with culture. Some cultures tend to be monochronic (for example, the dominant culture in the US), while some follow a polychronic time orientation (for example, those of many African and Latin American countries), in which human interactions are considered more important than time. Growing up in Germany, you can guess where my time orientation falls on that continuum. Of course, I was taught to think of time as a precious resource and to focus on completing every task at hand with utmost efficiency. Psychologist Robert V. Levine, who wrote the book *A Geography of Time*, would have probably called me the stereotypical example of someone who has always focused on "clock time."[153] In contrast, polychronic cultures live by "event time," in which an Internet session stops when external influences dictate it, not when an hour has passed or all to-dos have been completed.

Time and Decision-Making

Several years after my discovery of these cultural differences in time, I met Myke Naef—founder and former CEO of the online calendar and event scheduling tool Doodle—at a public research symposium in Switzerland. At that time, Doodle was already a leading online tool for scheduling meetings and was used in many countries around the world. We started chatting about potential cultural differences one could likely observe in Doodle users. And as so often happens with these seemingly random ideas, born was a new research project. In the subsequent months, I worked with my colleagues to analyze whether Hall's and Levine's theories about the differences in social time could be seen in the Doodle data logs. Together with his collaborator Norenzayan, Levine had compiled a ranking of the "pace of life" in 31 countries, produced using measurements of walking speed, postal speed, and clock accuracy in them.[154] In a nutshell, they had sent someone to different cities in each of these countries, asked them to record how fast people walked, how long it took to buy a stamp, and how precisely public clocks were set, and then had sorted those outcomes by country. Unsurprisingly, Switzerland ranked at the top of the list, followed closely by Ireland and Germany. Brazil, India, and Mexico had the slowest pace of life in their study. (Their data was collected between 1992 and 1995, so take this ranking with a grain of salt.)

Could we see these differences in time perception on Doodle? After a few months of developing and testing hypotheses on the Doodle data, we had the answer. Sure enough, Levine and Norenzayan's measurements were highly correlated with how often people in different countries use Doodle (as measured by the number of Doodle polls per Internet user).[228] Liechtenstein, Switzerland, Austria, and Germany were leading the list of 211 countries and territories that use Doodle, as if they had waited their whole lives for a tool that could make things even more efficient. People in countries that tend to run on clock time also generally initiated Doodle polls further ahead of time than those in event-time countries: Switzerland and Germany created polls on average more than 30 days ahead of time, while others, such as Vietnam, Chile, and Lebanon, were much more spontaneous, each with less than 8 days' advance scheduling. That's a difference of 22 days!

Of course, we cannot assume that everyone in these countries uses time similarly. For example, New Zealand was one of the countries that created polls far ahead of the actual event (on average 21 days in advance). But we also know from prior work that the indigenous Māori are more aligned with event time,[167] so they may want to plan more spontaneously. Unfortunately, the Doodle data only allowed us to compare countries, so it's hard to know whether the Māori use Doodle differently than the average Doodle user in New Zealand.

When digging in a bit further, it was striking to see how online scheduling reflects known social dynamics of groups of people. You may have experienced this yourself if you've used a scheduling tool like Doodle in the past. Did you generously indicate that most times would work, including options less convenient for you? Or did you try to protect most of your available slots? And did you find yourself looking at the options provided by others before deciding how many of them to agree to? Psychologists call the way people tend to learn what is appropriate from other people's actions 'social proof'. What we saw in the Doodle data was that the responses of earlier poll participants determine the responses of later ones.[231] But are the responses dependent on each other because of social proof? Convenient for us, Doodle also has "hidden" polls, in which respondents cannot see what others chose. This gave us a nice option for comparison: In hidden polls, we did not see that people's decisions were influenced by those answering the poll before them. If people's responses depend on each other's in open polls, but not in hidden polls, it seems clear that people are paying attention to others' availabilities and that this influences what options they choose themselves. And this is great! Doodle wants to support groups in coordinating and ultimately finding an option that works for all or at least for most, and their open polls do exactly that.

Unfortunately, this way of making decisions in a group doesn't seem to work equally well around the world. You can probably imagine where I'm going with this. We now know that people try hard to help their larger group reach an agreement. By doing so, they follow social procedures. And, of course, we can assume that these social procedures differ across countries and cultures. When we looked at polls by country, we saw that open polls in India find a much higher number of mutually agreeable options—28%—than any of the other 43 countries for which we had enough polls for comparison.[228] You can see the list of countries in Figure 4.1. At the far end is the US, where Doodle users generally find consensus on only 15% of options. Across all countries, we saw that Doodle users in more collectivist countries, such

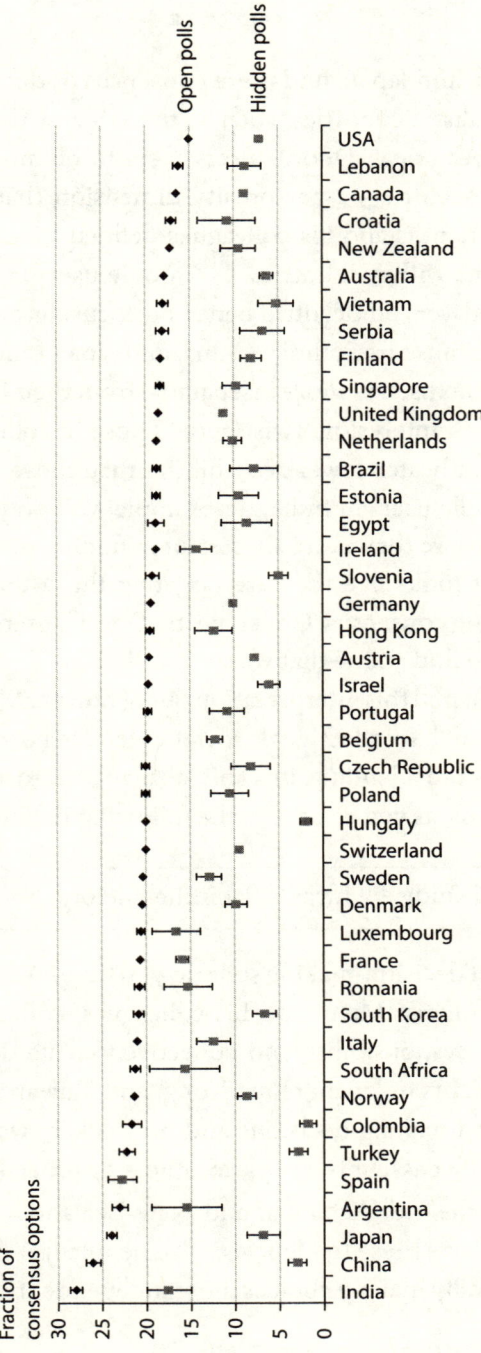

FIGURE 4.1. The average fraction of options, such as given dates and times, for which groups found consensus. Error bars show the standard error. Figure from Reinecke et al.[228]

as India, China, and Japan, find more consensus options than those in more individualistic countries, such as the US and Canada. In other words, North American Doodle users seem to be mostly idiocentric rather than allocentric—a personality dimension that cross-cultural psychologist Triandis and his colleagues defined to account for individual variations within countries.[281] Doodle users from the US and Canada showed very idiocentric behavior, focusing on their individual goals, while those from India, China, and Japan (and many others) seemed to see themselves more inseparable from their ingroup's goals.

But there is an interesting twist here: Those in collectivistic countries generally indicated fewer availabilities than those in individualist countries. Doodle users in India, for example, will only agree to a few options. So why are they more successful in finding options that work for all? Well, it looks like they are targeting those that have a good chance of finding consensus. Collectivistic Doodle users seem to make a larger effort to find a time that works for all.

We didn't just pull this interpretation out of thin air. Many anthropologists have stated that collectivistic societies tend to be more concerned with harmony and blending in. This also relates to how we prefer making decisions, as you can see in the following LabintheWild test:

LabintheWild study #3: https://labinthewild.org/bookstudy3

What kind of decision-making style do you have? According to prior work by psychologist Mann and his colleagues, collectivists tend to prefer to share responsibility and collectively make decisions. They found that students in Japan, Hong Kong, and Taiwan reported being less confident in making decisions and more likely wanting to avoid them (such as by passing the buck to others or procrastinating) than students from the US, Australia, and New Zealand.[175] How people use Doodle reflects these tendencies, with people from individualistic countries generally making choices more independently of others than collectivists.

To me, one of the most insightful takeaways from our Doodle study is that people around the world often use the same technology very differently. Despite all being Internet users who used Doodle for scheduling, we saw clear differences in their behaviors. Contrast this with an article published by Harvard economist Theodore Levitt in 1983, 30 years before we published our Doodle work. In his article, Levitt described how globalization and the availability of products worldwide has resulted in a homogenization of people's cultures: "Different cultural preferences, national tastes and standards, and business institutions are vestiges of the past."[155, p.98] Our Doodle study clearly showed that this assumption didn't quite pan out.

Dealing with Choice and Agency

The differences in people's behaviors on Doodle also suggest that people deal more or less well with a large number of choices. Some clearly prefer following others' choices, whereas others prefer autonomy in their decision-making. They prefer to have *agency*, as social scientists would say. But it's not always about preferences: People's agency in making choices is often limited by social structures that can (subconsciously) limit someone's ability to independently make choices. In fact, researchers have shown that Indian participants were less likely than American participants to understand their actions as choices, suggesting that seeing agency as something that is driven by personal preferences and goals is a very "middle-class European American" view.[242] People from India tend to see actions less as a choice, but instead as a way to respond to social roles and the expectations of others.

The interesting part here is that we did not always know this. For many decades, psychology research has led us to believe that people, in general, will understand their actions as making choices and that giving people choices is a good thing. Think about how most online stores are designed today and you may feel immediately overwhelmed by the number of choices they provide. Or think about how the Internet is designed with its numerous hyperlinks that require you to make choice after choice after choice.

The thing is, for most US Americans, having a choice equals freedom. It's a moral foundation established in the nation's Declaration of Independence and something that percolates through every aspect of American life, from educational practices to caregiving and interpersonal relationships.[180] Having autonomy and choice means being independent, which in turn suggests maturity (a child gets taught to become independent) and well-being (people can pursue what best satisfies their own preferences). It is now well established, across various tasks and domains in psychology research, that giving Americans a choice improves their personal control and intrinsic motivation to complete a task.[239] It also leaves them as more satisfied with the final product—I've secretly tested this on my American husband, who enjoys his cereal in the morning much more if he gets to choose it from at least five different kinds. Anything less, and there's trouble. The corporate world is well aware of this and readily provides a myriad of choices for seemingly simple things like buying cereal or ordering coffee. Do you want whole, 2%, almond, or oat milk? Which size? Iced or warm? Do you want it to go? Do you want to go through the painful experience of spelling out your name or would you rather use a fake name? While I certainly feel overwhelmed by the dizzying number of choices (which I've come to manage by always ordering the same under my dedicated "Starbucks name"), there is ample literature suggesting that this is, indeed, what makes the average American feel good about what they're getting in return. As Hazel Rose Markus and Barry Schwartz write in the *Journal of Consumer Research*, making these choices lets Americans form their preferences, express themselves, and experience control over their own destinies.[180] Choosing your own individual coffee can be a form of freedom.

But, of course, not everyone sees choice as a positive thing. And I don't just mean *too much* choice, as in when we experience choice overload. Because when we do, it can certainly paralyze us and make us dissatisfied with our decisions, no matter where we're from. What I mean instead is that having a reasonable number of choices doesn't have the same necessity around the world.

To illustrate this, let me tell you about a study by Heejung Kim and Hazel Rose Markus.[142] The two researchers recruited 27 Americans and 29 East Asians (from China, Taiwan, Hong Kong, and South

Korea) at San Francisco International Airport, asking them to fill out a survey. But while the researchers pretended to be interested in collecting survey responses, they were actually interested in something else: When offered a small reward for filling out the survey, a pen, which one would people choose? Once people had completed the survey, they were offered a choice of five pens. These pens were not chosen randomly—the researcher made sure to offer at least one pen with a different color from the rest. Which ones do you think people chose? Maybe you are like me to believe that people would choose their reward pen either randomly or based on their color preferences. But this is not what happened. Instead, a vast majority of Americans (74%) chose a pen that was different from the rest. It could have been the ugliest color, as long as it stood out from the others. In contrast, only 24% of East Asians did the same, meaning that the majority of them chose the more common pen.

I think this result is a great example of the impact that culture can have on the seemingly trivial task of making a choice. As Kim and Markus write in a summary of their results: "Where Americans preferred uniqueness, East Asians preferred conformity, and these preferences were associated with divergent individual actions."[142] Not everyone will strictly fit into these cultural molds—as you can see from the results, there were several East Asian participants (24%) who chose the unique pen and several Americans (26%) who chose the common pen. The choices we make are also arguably influenced by the situation we are in and many other factors. But it is difficult to deny that culture influences how we choose.

There's one additional question we could ask ourselves: Do we equally benefit from following others' choices and independently making them? Or do we perceive one or the other as more satisfying? (After all, it can be quite a burden when you simply want to grab a coffee!) To answer these questions, consider participating in a computer game. Most of today's games take a ridiculously long time to set up because they offer so many options for personalizing your player name, character avatar, and other things that are considered relevant to your self-expression and enjoyment of the game. But once you've completed the setup, do you think the choices you've been given will impact your motivation or even your performance in the game? In a 1999 paper in the *Journal of Personality and Social Psychology*, Sheena Iyengar and Mark Lepper reflected on the conventional assumption that people universally find being given a choice intrinsically motivating.[132] "But are these principles truly as self-evident and as universal as they might first appear to investigators raised and living in North America?" they wondered. In short, they didn't buy it. So they set out to do several studies that compared whether being given a choice is equally intrinsically motivating for people from different countries. In one of the studies described in their paper, Iyengar and Lepper asked fifth graders from the San Francisco Bay Area to play an educational game, Space Quest, which was designed to teach basic arithmetic equations. Roughly a third of the students were given a version of Space Quest in which the first step was to choose a spaceship avatar and a display name. This was the *personal choice* condition. The remaining students were put into one of two conditions that didn't allow them to choose the avatar and name. Instead, they were assigned to either an *outgroup condition* (in which they were told that third graders at another school—the outgroup—had preselected what the spaceship will look like and what the crew calls them) or an *ingroup condition* (in which they were told that the choice of the preselected spaceship and name was informed by the preferences of the majority of other students in their class, their ingroup). Of course, the students were unaware of the experiment conditions and what the researchers were testing. They were simply told to play the game and learn what they could, and that they would not be graded on this.

Here's what happened. European-American students in the personal choice condition played 4.7 rounds of the game, on average, but less than 3 rounds in the other two conditions. In other words, their motivation to play the game was negatively impacted if others chose the avatar and name of the spaceship for them, no matter whether those choices were informed by the preferences of an outgroup or an ingroup. But the results were almost flipped for Asian-American students. In the personal choice condition, Asian-Americans played only 3.7 rounds of the game, so approximately one whole round less than the European-American students. When the choices were made by an outgroup, Asian-American students played an average of 2.6 rounds of the game. But they played 4.9 rounds, on average, in the ingroup condition, seemingly feeling motivated by the idea that their spaceship was chosen based on the majority preferences of other students in their class. Moreover, these findings were also reflected in their personal liking of the game. When the researchers asked each student, "How much would you like to play the Space Quest math game again?," European-Americans in the personal choice condition reported liking the game most, whereas Asian-American students liked it most in the ingroup condition.

You have to admit that these are really quite stunning results. At the time the paper came out, in 1999, it resulted in a bit of an awakening for many psychology researchers, most of whom were Americans or at least resided in the US. Being given a choice does not always produce psychological benefits? And living with the choices of others is not always detrimental? It was hard to believe. But it's also not too surprising given that the Western view of individuals as independent and self-determining beings is not universally shared.

In an article published in the journal *Psychological Review* in 1991, Hazel Rose Markus and Shinobu Kitayama had already speculated that East Asians and Westerners tend to differ in how they think of themselves, how they have "different construals of themselves." They described East Asian cultures, such as Japanese, as *interdependent*, in that people foreground relationships and connectedness with others: "Within such a construal, the self becomes most meaningful and complete when it is cast in the appropriate social relationship."[178, p.227] In Western cultures, they wrote, people tend to have a more *independent* view of the self and strive to promote their own goals, be unique, and, well, be more independent from the rest of society. People who are interdependent share the responsibility for each other's well-being and behavior. Whereas independent societies value one's ability to express oneself as an individual, interdependent societies place more importance on nurturing relationships and adhering to specific roles and norms. This also means that a person who insists on their preferences and personal choices is considered immature in interdependent cultures because their insistence doesn't allow them to adjust to others.[180] I think this also shows that choosing and expressing our own preferences, as Westerners often strive to do, isn't always the best thing; it doesn't readily imply freedom and well-being, as the assumption often goes.

Shinobu Kitayama and others have come up with a neat little test to see whether someone has a more independent or interdependent view of themselves.[146] Please take it for a spin on LabintheWild.

LabintheWild study #4: https://labinthewild.org/bookstudy4

If you completed the test, you might have been surprised to see how such a simple test could give you so much information about yourself. Having you draw a diagram of your social network and inferring your cultural tendencies may even feel a bit like magic, but I think that's exactly the beauty of this task. In fact, I love this test for its simplicity and for not relying on survey questions (or self-report scales, as researchers call them), which aren't so reliable for comparisons of cultures.

Did you draw your own circle in the middle and situate everyone else around you? And did you subconsciously draw your own circle much bigger than those of others? If yes, you probably got told on the results page that you have an independent self-construal. If you drew the circles similar in size to those of others, you may instead have a more interdependent self-construal.

Kitayama and his colleagues used this test to assess symbolic self-inflation.[146] If someone sees themselves as mostly independent of others, the research team hypothesized, they will be more likely to self-inflate their representation in the sociogram task by drawing a bigger circle for themselves than someone who sees themselves as interdependent. The researchers asked college students from university campuses in Japan, Germany, the UK, and the US to draw a social network of their friends on a piece of paper and then measured how much larger the self-circle was in comparison to the average circle of the participant's friends. A smaller number means there is less of a difference between the circle sizes (interdependent), while a larger number means the self-circle was made much larger (independent).

Their results showed that the American, German, and British students all drew their own circle bigger than they drew others'. But not so the Japanese: Their self-circles were usually the same size as the circles of their friends, or even a little smaller. Of all four countries, the US participants showed the most independent self-construal, whereas Japanese participants were the most interdependent.

A study led by Thomas Talhelm later repeated the task in China, showing that Chinese participants, like the Japanese, are mostly interdependent.[266] But there was an interesting surprise in their finding—one that reminds us to not assume that people within artificially drawn country borders behave and think the same. What Talhelm and his colleagues found is that it matters whether their participants had been raised in the south part of China, which has a tradition of rice farming, or in the wheat-growing north. You can see the areas that are more engaged in paddy rice farming and in wheat farming in Figure 4.2. To rule out potential effects of dialect, climate, and exposure to herding, the authors recruited participants from neighboring counties along China's rice-wheat border, where you see the transition from black to light gray. Interestingly, people from wheat-growing areas drew their self-circle slightly bigger than their friends' circles, while those in rice-growing areas made them all roughly the same size. Why is this? The authors suggest that traditional rice farming regions developed more interdependent self-construals because rice farming requires a lot of coordination: From planting the rice to irrigation, farmers have to agree and collaborate with their neighbors and villages to manage water use and build irrigation networks. In contrast, wheat farming takes only half as much work as rice farming and is often managed by individual farmers. Across all these studies, the differences in the average self-inflation are not immense—have a look at Figure 4.3 to see the results of these studies in comparison. Nevertheless, I think it's striking that there are any differences at all in how people perceive the importance of themselves relative to their social network. It's something that explains the results of our Doodle study, and so much more.

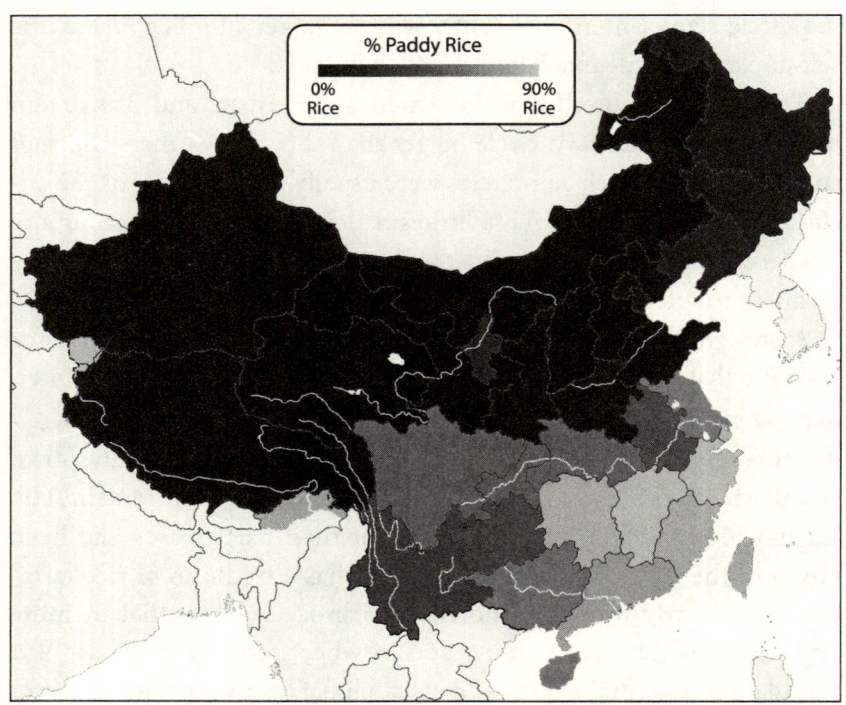

FIGURE 4.2. Distribution of paddy rice farming across provinces in China. Rice paddy data is the percentage of farmland devoted to paddy rice, from the 1996 China Statistical Yearbook. Image courtesy of Thomas Talhelm.

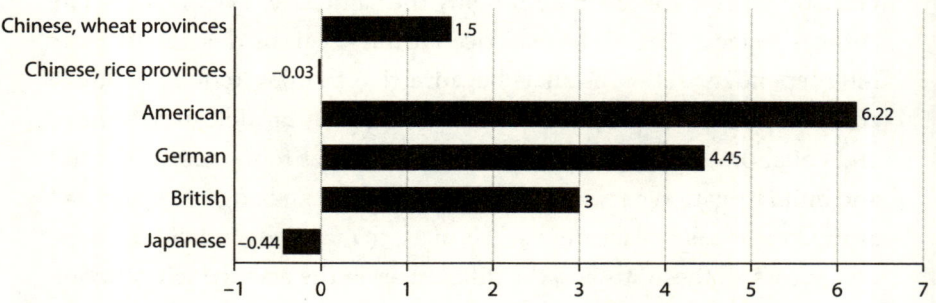

FIGURE 4.3. Results from the sociogram test have shown that different groups of people draw a circle representing themselves bigger or smaller than the circles representing others—the difference is shown in millimeters based on studies by Kitayama et al.[146] and Talhelm et al.[266]

Choice Overload in Navigating the Web

What do the findings about people's preferences for more or less choices mean for design? Years of research have exposed that having too many choices isn't good for anyone. Overwhelmed by a large number of complex choices, people just give up. But choice deprivation isn't great either: people are generally less satisfied with not having any choice than with having too much choice, though the happy optimum is likely determined by individual and cultural preferences.[230] The thing is, the online world has resulted in a sheer explosion in the number of choices compared to what we're commonly used to in our offline lives. My local bookstore has books stacked all the way to the ceiling, but the number of books it was able to cram in was ultimately constrained by the limited shelf space. Online, shelf space is suddenly unlimited. Of course, this can be great if you are looking for a very specific book. But I think it's fair to say that if you're simply browsing and wanting to find a new book to read, you will prefer less of a choice.

This is where recommendations come into play. Amazon, the Chinese Taobao, Netflix, YouTube ... they all are keenly aware of the problem of choice overload. To help you choose—and make more money—they tell you what others have chosen. Some of the credit for really popularizing these recommendations goes to Amazon and its shopping cart recommendations.[160] The idea was: When people put something into the cart, let's show them other products based on what they are about to purchase. In 2013, a McKinsey report found that 35% of the products people purchase on Amazon and 75% of what they watch on Netflix are chosen based on the recommendations.[172] Needless to say, it's a huge success for companies if people buy more and spend more time on their platforms.

Let's think back to the idea that some people are more independent while others are more interdependent. According to the results of the Space Quest study, we could assume that interdependent people are more likely to be satisfied with recommendations from members of their ingroup, such as from their family, friends, or classmates. This certainly seems true for Chinese consumers, who tend to be more trusting and willing to buy something when it was recommended by an ingroup rather than an outgroup.[79] They are also more likely to seek out those recommendations: Researchers found that people in more interdependent, collectivistic societies, such as China, tend to use social media to help make purchase decisions substantially more often (for more than 60% of their purchases) than those in more independent, individualistic societies, where people only ask their online social network for suggestions for less than 10% of their purchases.[102]

But let's hit the pause button for a moment. Did you notice that this is not how online stores currently try to help you make a choice? Product recommendations do not show what your social network, your family, friends, or classmates, would choose. Instead, they show you what an anonymous set of customers would choose. By doing so, companies may be working with the data they have—after all, we can all be grateful if they don't also know who your friends and family are. (Or at least that they pretend they don't.) But, more likely, I think this is a design decision that seems perfectly reasonable to Americans, for whom making autonomous choices, independently of their ingroup, is often more standard. Put differently, there are default choices to design that are rooted in how technology design started and has evolved, and these default design choices commonly ignore the diversity of people.

There is another fundamental difference between cultures that could impact how we benefit from product recommendations: The extent to which we are motivated to fit in or to stick out. Think back to the study where participants were offered the choice of a pen as a reward for completing a survey.[142] If Americans more often choose a unique pen than one of the pens with a common color, wouldn't you assume that they also prefer outlier products in other scenarios? But, of course, choices aren't always about products. Most technology products today offer a myriad of options for personalizing everything, from avatars to skins that you can "slap on" to fashionably dress a user interface. And even navigating the Web is a constant exercise in making choices and not regretting them. Which link in the search engine should you click on? Which menu item should you choose first? In *The Smarter*

Screen, UCLA professor Shlomo Benartzi calls this experience "navigation overload."[35] While we can probably all relate to the feeling of being lost in hyperspace, researchers have long suggested that this experience may be worse for people used to fewer choices. For example, Hofstede suggested that there are cultural differences in teaching and learning that could determine how comfortable we are with freedom of choice and unstructured environments.[125] In societies with a low power distance, say, Denmark, with its power distance score of 18, or the US, with a score of 40, teachers strive to provide a student-centered education, valuing student initiative and expecting them to find their own learning paths. A lot of emphasis is on self-directed learning. Learn how to learn. Hofstede described how the opposite is often true in societies with a high power distance, where rote learning can be more common. Malaysia is one example, with a power distance score of 100, and so is China, at 80. Hofstede also suggested that uncertainty avoidance is related to this, his cultural dimension that describes people's tolerance for choices and ambiguity. According to Hofstede, having many choices is fine in cultures with a low uncertainty avoidance, such as Singapore, Jamaica, Denmark, and Sweden (the four countries Hofstede found to have the lowest uncertainty avoidance). Members of those cultures often feel fine navigating ambiguous situations. But in countries with high uncertainty avoidance (Greece, Portugal, Guatemala, and Uruguay top the list), people feel more threatened by unknown situations and expect more predictability. Hofstede speculated that students from high uncertainty avoidance cultures are likely more comfortable with structured learning situations, such as those that provide detailed objectives and assignments. In societies with generally low uncertainty avoidance, having vague learning objectives and assignments, and generally a more unstructured learning environment, might be more acceptable.

If this is true, we could expect that some people are indeed more familiar and comfortable with unstructured, nonlinear environments on the Web than others. Could we see this in how people navigate web pages? In 2005, a team of researchers from Humboldt University and Free University Berlin found some evidence for this.[148] The team analyzed data from a large, international public health website that had been translated into English, German, Spanish, and Portuguese and visited by users from 187 countries. Among other results, they found a difference across countries in how many pages people access and how linear their navigation behavior is. Accessing one page after the other was considered linear, whereas backtracking to a previously viewed page was considered nonlinear. The team found that this behavior correlates with uncertainty avoidance: The higher the uncertainty avoidance score of a country, the higher the number of pages people from that country accessed. Perhaps covering most information (often in a linear way by almost systematically going from one page to the other) is important for people in high uncertainty avoidance countries to reduce the risk of the unknown. The researchers concluded that having many choices—many paths to explore—is beneficial for low uncertainty avoidance cultures, but should be avoided when designing for high uncertainty avoidance cultures.

I remember having great discussions with Philip Guo, a Human-Computer Interaction researcher and an expert on learning tools, about the implications of such cultural differences for the design of MOOCs, the Massive Open Online Courses that suddenly began to emerge out of nowhere in 2011 and 2012. They generated huge hype. Coursera, Udacity, and edX are all examples of MOOC providers that were founded at this time by US scientists. (Since then, country-specific platforms have been launched in many different countries, from India's Swayam to Jordan's Edraak.) Philip and I were both in Boston at the time—he was at MIT and I was at Harvard—so it was easy to meet and speculate about the future of technology and its impact on society. And Philip had signed up for a summer gig as a visiting research scientist at edX, a MOOC provider that had been founded by MIT and Harvard scientists in 2012, so many of our discussions naturally focused on this new phenomenon of MOOCs. Like Coursera and Udacity, edX offers free online classes that anyone can take. It was one of those moments in history when many people, myself included, got very excited about the impact that education could have on people's lives around the world. "What if everyone had access to course content that Harvard and MIT usually only offered for a big chunk of change?" the thought was. People hailed this development as a way to democratize education—a pathway to reducing poverty in developing nations.

The only issue was that nobody thought deeply about how to present the course contents. There was already much literature on individual differences in learning styles—scientific findings that teachers have increasingly used to offer more personalized education in the classroom. And we also already knew that there are vast cultural differences in how people around the world provide and experience education. But when MOOC platforms were designed, most of this was ignored. Give people access to all course content and they will be able to personalize their education themselves, the designs of these platforms seemed to imply.

Indeed, MOOCs are often designed around the notion of freedom of choice. They do have some structure to them; for example, each course is usually subdivided into a series of weeks, with each week containing several web pages that include video lectures and/or assessment problems. But unlike in a physical classroom, there is no oversight in how students access the content. You don't "unlock" more learning content once you've completed others. And nobody will stop you from starting at the very end of a class and working your way backward, or skipping most of the content. Even the assessments are often viewable at all times, which means you can easily cheat by looking at the questions first and then finding the content that helps you answer them. In other words, you have to be quite self-disciplined to learn for the sake of learning and to do this without a teacher who could provide you with guidance and exercise authority. And you have to be okay with the numerous choices MOOCs offer: Should I follow its structure from beginning to end? Should I view the video first? Or start by taking a peek at the assessments?

Philip and I wondered how this kind of "freedom" might play out for the diverse group of online students on edX. We analyzed the data from four of its MOOCs, which at the time had been accessed by 140,546 students from 196 countries.[107] You can see from these numbers that MOOCs can have quite an impressive reach. Many people around the world access their content, and most of them are probably set on learning from it. But the diversity is skewed. Across the four courses we looked at, most students who ended up getting a certificate for successful course completion came from the US (19%) and India (12%), with all other countries making up 8% or less of the certificate earners. (Language might be an issue here, but Spain, Russia, and Germany still had significant numbers of learners in the four courses despite English not being one of their official languages.)

Motivations for accessing MOOCs can of course vary. Not all people use it to gain knowledge and skills, though this is often the primary reason.[44] Many also take part because they see it as a personal challenge ("Could I pass a class taught at MIT?").[44] You can also imagine that not everyone plans on completing the MOOC and earning a certificate at the end; some people may be perfectly happy learning for their own benefit or are simply curious about how the course content is taught. But we could assume that those MOOC students who earned a certificate shared the motivation of earning a passing grade. Any variations in how these certificate earners used a MOOC should be less likely due to differences in motivation.

So Philip and I started comparing the behavior of certificate earners across countries. In what order did they access the learning content? How much of the learning content did they cover? It turned out there were fundamental differences across countries in how students went through the MOOCs. For starters, students varied in how much of the learning content they covered, which signals that some were doing the minimum required to earn a passing grade while others may have been more intrinsically motivated to advance their knowledge. Indian certificate earners in our dataset, for instance, covered 71% of the content, on average, while US certificate earners covered 83%. Perhaps unsurprisingly, almost nobody looked at all course materials. In fact, the average certificate earner in our dataset never even accessed 22% of the course content, yet they all passed the course. But how much content they accessed mattered: We found that the more course content someone looked at, the higher their grade was. So even though we don't know how deeply they covered a given page or whether they learned anything from it at all, accessing more pages resulted in better grades.

Of course, this means students had to make many choices about which parts of the course to access and how much to cover. They also had to make choices about the order in which to do so. And this is where it became really interesting. When Philip and I looked at the data by country, we found that students in some countries had a much more nonlinear navigation strategy than others. Specifically, we analyzed how many times students navigated backward in the course (say, from Lecture 7 to Lecture 3) relative to the course content they covered overall. Almost everyone does this, jumping back from a lecture later in the course to one of the previous lectures. But overall, students from Nigeria, Kenya, India, and Pakistan went through the learning materials fairly linearly, often closely following the given path. They still jumped back to already visited lectures, but they did so much less often than students from the US, Russia, and many European countries. Those students often jumped back at least once per visited page, overall navigating the content in a highly nonlinear way. In other words, they were making lots of choices in their learning strategy. Psychologist Herman Witkin would have described them as *field-independent learners*[313] —people who are confident in freely exploring the learning content and defining their learning paths. Witkin and other researchers contrasted this cognitive style with *field-dependent learners*, who usually prefer following the path defined by a teacher or a learning environment.[84,158]

We already had a hunch why we saw these differences across countries. After all, there are lots of differences in how children are educated across countries, and this has a tremendous influence on their learning style.[125] How their classroom interaction is structured depends on the values of their society. But it is also influenced by the financial resources that schools have available. Rich countries often have relatively small class sizes (a *low student-teacher ratio*) because they can pay for more teachers, classrooms, and other resources. Poorer countries do not always have this luxury. When Hofstede wrote about these differences in 1986, he proposed that students in countries with a high student-teacher ratio may be more used to a teacher-centered education, observing and following the lectures rather than actively defining their own path.[125]

Think about how you have experienced education. How many students did your teachers have to oversee in elementary school? Were the lessons interactive, with students frequently taking the role of question-askers and hands-on activities being the norm? Or were you mostly asked to listen to the teacher and answer questions when you were asked? If you experienced the latter, a fairly teacher-centered education, you may have also experienced teachers going through the learning content mostly linearly, covering one topic after the other, and perhaps requiring you to memorize facts more than focusing on understanding the bigger picture. I certainly had that experience in some of my classes in high school and college. But in other classes, I noticed my teachers were trying to point out the relationship between the different topics they taught us. They would go out of their way to draw connections between a historic event we talked about in Week 7 and an economic crisis that was discussed in Week 4, for instance. In other words, they were jumping back and forth between previous and current lectures to tie the content together. Could this experience make people more field-independent and self-directed when learning with MOOCs?

Indeed, when we compared the navigation behavior of the MOOC students with student-teacher ratios in primary schools (data that is collected and made available by the UNESCO Institute for Statistics), we noticed a decent correlation between the two variables. The higher the student-teacher ratio, the more linearly students navigated through the MOOC (though they left out more content than others). According to the UNESCO data, the countries whose MOOC users navigated the most linearly, such as Nigeria, Kenya, India, and Pakistan, all have between 30 and 44 students per teacher in an average school. In contrast, those that jumped back many times (the US, Russia, and European countries) have at most 21 students per teacher.

There's of course a big caveat in drawing conclusions here because these student-teacher ratios are averages. Many countries have huge variations in these ratios, sometimes depending on public and private schools, or depending on local laws. We have no idea what class sizes the MOOC students are used to and how this may have translated into a more or less teacher-centered education. That is to say, the MOOC users may not be representative of the average population in their countries. Yet, intuitively, our finding seems to make sense. Countries like Nigeria, Kenya, India, and Pakistan are considered societies with a high power distance. Rote learning and following strict instructions are not uncommon in these countries, partly because of larger class sizes.[125] So it is entirely plausible that being used to larger class sizes, and the more teacher-centered education that often comes with them, trains students to become more linear learners. In the absence of a human teacher, students from these countries tend to follow the structure predefined by the MOOC.

And this may be a good thing—after all, the MOOC providers must have thought about how to best offer their content. Except we also saw that students who followed the linear structure were less likely to earn certificates. The difference was immense: Those who did not earn certificates only jumped back an average of 0.3 times per visited page, while certificate earners did an average of 1.04 jumps back for every page they visited. Jumping back and a nonlinear learning strategy seem to be key to success in MOOCs. In addition, linear learners who earned a certificate skipped a lot of content. They covered less content overall than the nonlinear learners. And this meant that, overall, their grades were lower. Put simply, the students who employed a "more linear, but lower coverage" strategy ended up learning less.

In the weeks after our initial analysis, Philip and I had many discussions about the implications of these findings. What the data showed is that MOOCs can be great for people who are already used to self-directed learning. But not so much for others. While more research is needed (as we scientists like to say for anything that we do), US-designed MOOCs seem to be better suited for intrinsically motivated and self-directed learners than for those who only cover content that lets them barely pass the hurdle to get a certificate. In other words, the promise of democratizing education was flawed by a Western bias in design.

Since then, various country-specific MOOC platforms have been developed by governments and organizations around the world to serve new educational strategies or to compete with US MOOCs. And while we may have thought that the design of MOOCs was set in stone, some researchers have pointed out interesting differences between their local MOOCs and those designed in the US. For example, XuetangX, the first Chinese MOOC platform launched in 2013, includes bustling online spaces for students to interact with each other and to share self-produced content, such as videos about a specific topic.[248] In a paper led by Shuqing Liu from Tianjin University in China, her team compared the design and workflows of Coursera to the Chinese University MOOC platform, called iCourse.[166] I thought one of the findings was particularly interesting: Rather than organizing content according to a week-by-week timeline, as is done on Coursera (and edX, as we discussed above), the developers of iCourse organize content by categories, including lectures, assignments, videos, and exams. Liu and her co-authors show that Chinese participants expect to have a list of videos that is separate from other learning content and that this is preferred over mixing learning materials across categories. While the content within these categories is still sequential—for example, one would assume that people view the first video before the second—Liu and her co-authors describe the organization of the course content as a "parallel classification approach." This approach seems to also be more natural for their Chinese participants: Compared to working with Coursera, their participants felt more comfortable finding learning content in the parallel design.

Digital Utopianism

I'm glad to see these local MOOCs using different designs than what we became used to on Coursera, edX, or Udacity. I'm especially glad because some countries have opted to build their MOOC platforms on top of the open-source Open edX platform provided by the nonprofit platform Axim Collaborative[22] (though more than half of the platforms employ English, which severely restricts their utility to a small part of the world's population[6]). For example, IITBombayX, a non-profit MOOC platform that is popular in India, used Open edX.[53] It's great that Open edX provides this service; building on open-source code can be a huge cost saver for governments, and in some cases, it may be the only way to introduce a "local" MOOC platform in their country. But there's also a big problem with this. Using the Open edX code also means that many MOOCs will look like edX. Having a week-by-week timeline, for example, is already baked into the design. Researchers have also pointed out that many of the existing MOOC platforms are incompatible with the Internet infrastructure in many countries where low bandwidth and low computing power prevent access to their content.[53] Indian users perceived the MOOC forums as no better than "an information desk," without them providing a place for interacting with others.[195] So, overall, offering open-source MOOC platforms seems like a good thing. But it also promotes a one-size-fits-all design—despite our knowing that the local context and culture influence how people best learn and that even within countries there are substantial variations. (We will talk more about differences in relationship-building and social interactions in the next chapter, though you'll see that this will be a common thread throughout this book.)

Alas, I fear that MOOCs have revolutionized education for some, but not for all. I'd even go so far as to say that MOOCs provide the same kind of technological solutionism that Kentaro Toyama criticizes in his book *Geek Heresy*.[276] As a computer scientist and long-time Microsoft employee (before he joined the faculty at the University of Michigan, where I was fortunate to be his colleague, albeit only very briefly), Kentaro has spent much of his career believing in the power of technology. He took part in research projects motivated by the goal of reducing poverty and bringing education to all, such as by developing educational games that several children could interact with at the same time using multiple computer mice,[213] or by providing digital videos with agricultural information to farmers in India.[93] His belief was that technology can alleviate poverty.

But here's what I greatly admire about Kentaro: He continued to question his beliefs. Despite spending many years working for one of the largest and most prominently known tech companies in the world—a company that at its core values the transformational power of technology—he remained open to changing his viewpoints. And at some point, he came to the challenging conclusion that the belief in digital technology is a cult. It's a seemingly universal belief that technology will rescue our society from its ills. We rarely question whether digital technology has improved our lives. For many of us, it has indeed. But what Kentaro points out in his book is that our digital utopianism is often unsupported.

Since we were talking about MOOCs, consider education as an example. As Kentaro describes in his book, there have been a large number of technology projects designed to support education in a quest to bring information to all. The One Laptop per Child Project is one example;[15] it was inspired by the deep belief that a robust, low-cost, and low-power laptop could provide computer literacy and access to information that would ultimately lift children in the developing world out of poverty. Many projects around computer-based instruction had similar goals, the common idea being that digital education is a solution to the woes of learning. I firmly believed in this idea myself—my 2005 stay in Rwanda was meant to help develop educational software for small-scale farmers, and, motivated by this project, I later started my PhD with a project on e-learning. Digital education, I thought, could scale and bring knowledge to many more people around the world.

These beliefs are not entirely supported. Research has found over and over again that while computers and computer programs can be beneficial when *supplementing* good education, they can't replace good teachers. And, as we've seen, this is also true for MOOCs. They are great for some people, though they still can't replace good teachers. But they especially fail to adequately support those who are less adept at self-directed learning.

Ultimately, MOOCs could even increase the digital divide. Kentaro calls this the "Law of Amplification"—those who already know how to learn by themselves will be amplified by having access to MOOCs. They will get more education and grow in their knowledge and perhaps even in their careers. Others will not benefit as much even if they have access to MOOCs, or they may not benefit at all.

I think this example nicely illustrates what happens if we rely on one-size-fits-all designs, as is so often the case. Sometimes it means that technology is not as usable, useful, appropriate, or engaging for people because of differences in time perception, decision-making, choice, and agency, or differences in independence and interdependence. In the worst case, designs imposed on other countries and cultures (most often through Western technology products) can increase the digital divide: some people will be amplified by them if they adequately support them in their goals, while others will not have the same benefit. So unless we find out how people differ in their use of technology and how we can amplify everyone, no matter their cultural background, local context, learning style, or individual preferences, we risk increasing inequality.

5

Use of Online Communities across Cultures

If the idea of clock time versus event time from the previous chapter is still in your head, you're not alone. It's a fascinating way of thinking about our relationship with time and how it is governed by cultural norms. It's sometimes shocking to me (and to people around me!) how I'm so clearly in camp clock time. I wake up and eat at roughly the same time every day and my days are usually filled with meetings and lectures that start and end at a specific time. I also have the annoying habit of arriving too early to any meeting, which is a waste of my own time and creates a weird dynamic when I meet with friends and colleagues who are regularly late. In short, I find it difficult to ignore the clocks around me—and yes, I even find this difficult when I'm on vacation. And it's not just me: clock time is the dominant mode in most Western countries. In some it is treated almost like a religion. There are also many technology products that promote clock time. Unless you have a paid plan, Zoom (the video conferencing system) shuts off after 30 minutes no matter whether you've said goodbye or not. Calendar software rigorously slots our days into pre-set 1-hour meetings unless you make tedious adjustments. Google Maps tells us exactly when to leave to be on time for a meeting across town. And so on.

But what is missing when we schedule our days around time? Event timers would say the missing piece is meaningful social interaction. People who tend to live on event time often prioritize their interactions with others over time and punctuality. They are also often more collectivistic; Triandis and his colleagues have found them to cherish establishing and maintaining relationships over efficiently completing a task.[281]

And this translates into how they use the Internet. A 2002 study led by Patrick Chau, at the University of Hong Kong at the time, showed that undergraduate business students in Hong Kong were more likely to use the Internet for social communication than an equivalent group of students in the US.[52] The US group was instead more likely to say they use the Internet to search for product-, education-, or work-related information. In another study from 2011,[320] researchers found cross-cultural differences in the importance of socializing even for users of social networking platforms (which I had assumed everyone uses for social purposes). What the study found is that Chinese and Indian users of social networking platforms post questions in their status updates more often than do users from the UK and the US. They were also more likely to engage in answering questions. The reason for this? According to the study's results, Chinese and Indians trust those answers because they have a relationship with the people they are connected with. To grow and maintain those relationships, they also make sure to answer the questions of others. Posting questions and answering those of others helps enhance social connections, and Asian users especially seem to make use of this.

The Exchange in Stack Exchange

Fast-forward to 2018, when my then-postdoc Nigini Oliveira set out to investigate how people across countries use online Question & Answer (Q&A) communities. You may have used one of them yourself before: sites like Quora, Yahoo! Answers, Zhihu (in China), or Stack Exchange (a network of Q&A sites on different topics). They all promise to help users find answers to their questions quickly and efficiently. But unlike social networking sites like Facebook (or the popular local equivalents Weibo in China and Orkut in India), Q&A sites are focused on, well, questions. And answers. They are usually less focused on the idea of forming deep connections with other people. In line with this, we found in our study that some of these sites almost aggressively promote individualistic values, such as productivity and efficiency, and clock time.[207] On Stack Exchange, a tour for newcomers makes this very clear by highlighting its mantra, "Ask Questions, Get Answers, No Distractions," and later explaining, "This site is all about getting answers. It's not a discussion forum. There's no chit-chat." These goals are consistently pursued throughout the design of the forum and its rules. For example, people are discouraged from thanking others when they answer their questions. Instead, Stack Exchange encourages users to upvote the answer as a way of making them float to the top to help others more efficiently find good answers. Strict rules also underlie adding comments to a post: "Comments are meant for requesting clarification, leaving constructive criticism, or adding relevant but minor additional information—not for socializing."

For someone operating on clock time, this can be great. You ask a question, quickly get an answer . . . and, boom, you can go on with your life. Even more efficient is browsing other people's questions and finding the best answer (or at least what is considered "best" by other users). It's super time-effective, so you can quickly get back to your task and meet your deadline. But what if you are less interested in optimizing for efficiency and more interested in social interactions? Event timers have a more flexible relationship with time because their cultural background often taught them to emphasize interpersonal, collectivist aspects. When Nigini and I were discussing this project, he suggested that productivity may mean something different for collectivist users. As a self-identified event timer, he felt that finding an answer to a problem on these online Q&A sites may satisfy people's information needs, but that doesn't mean that collectivistic users would truly feel part of the community.

The data confirmed his suspicions: In our interviews, Chinese and Indian participants seemed baffled by the lack of interpersonal exchanges on a site that calls itself Stack Exchange. They described their goals for using these sites as a way of socializing, or "to relax and have fun," as one of the participants said. Because there is "no fun mode" in

Stack Exchange, the Chinese participants in our study told us that they generally feel more comfortable being on Zhihu, the "Chinese Quora" that has been thriving since 2011. Of Zhihu, they said, "it's more about having a conversation than getting an answer that is right," and that there are more personal questions and discussions. Clearly, social communication in these online Q&A communities is an important goal for Chinese and Indian users.

This is highly relevant for platform designers. Online communities rely on voluntary contributions of people's time and effort, which are needed to post questions, to provide answers, to up- and downvote content, and even to moderate content. So online communities only work if people are willing to actively contribute to them. And for this, people need to feel part of the community. The problem is that most online communities have many users, but only very few content contributors. In fact, in one of our studies, we compared the content contributions people make to Stack Exchange across 166 countries.[206] We looked at those who had created an account on Stack Exchange, assuming that these users would be more invested in the site and its community. But even among these account holders, we saw large differences in the number of contributions across countries. For example, almost half of the account holders from China and South Korea never contributed any content. In most Western countries, such as Germany, Switzerland, and the Netherlands, and also Israel, only around 20% didn't contribute— everyone else contributed to the community in some way or another, such as by posing or answering questions.

Now, it is of course much more difficult for people to feel empowered to contribute if they are not fluent in English. And I certainly think that plays a role here. One of the great barriers of the Internet is that much of its content is unavailable in languages other than English. But we found that language plays only a partial role in predicting how much someone will contribute. Whether a person is from a more individualistic or collectivistic country plays a much larger role according to our analysis. It seems that this is also consistent with other platforms. For example, the average person on Yahoo! Answers provides more answers if they are from a Western, individualistic country, such as France, Australia, or Germany, than if they are from a collectivistic country (e.g., Ecuador, Venezuela, and El Salvador).[138]

Zhihu, the Chinese Q&A platform I mentioned earlier, seems to get a bunch of things right when trying to engage its users and make them feel part of the community. As a user, you automatically get assigned a "salt value," which shows, well, how "salty" you are—the lower the score, the saltier. (I've also been told that "salt value"—or 盐值 yan zhi—is a homophone of 颜值, which is "beauty value," or how attractive someone is.) Some of the criteria that determine this score are how many details you have provided about your background and experiences and how much content you have contributed to the platform in the form of questions and answers. But, most importantly, your salt value increases the friendlier you are, the more you respect the rules of the platform, and the more you help govern the platform, such as by up- and downvoting content, or by editing and reporting questionable questions and answers.[290] Basically, the salt value encourages you to be a good community member. The platform also gives out colorful badges to encourage interactions among its users. If you access the platform every day, you'll soon get an "active user" badge, for example. If you follow many people on the platform, you can earn a "sociable person" badge. And if you contribute information in your field of expertise, you may get rewarded with an orange badge.[284] All of this is a way to make it easier for people to see who are the active and trustworthy members of the Zhihu community and to motivate them to do their own part. Chitchat encouraged.

Cultural Shaping of Online Spaces

What do people do when the online communities they'd like to use don't match their expectations? I think they have several options. One is that they find ones that are better suited to them, perhaps a local version like Zhihu, if one exists. Another option is that they continue using the suboptimal community, but probably only begrudgingly. As you saw earlier, there are many people who use Stack Exchange passively, such as to obtain information, but do not actively contribute content.

A third possibility is that they use and customize the space for their own needs. Silvia Lindtner, Ken Anderson, and Paul Dourish called this "cultural appropriation"—a term that has since evolved and changed in its meaning. In one of their articles, they described this phenomenon as when "information systems are often put to novel and unexpected uses, and they may be tweaked and transformed to achieve goals never imagined by their designers."[163] Users are not passive recipients of design decisions, but they actively shape them.

This active shaping of how technology products are used is a fascinating topic in the context of online social networking platforms. Many of the earlier, widely used social networking platforms were invented and designed in the US—think SixDegrees.com, Friendster, Myspace, Facebook, and others. While several of these platforms are now among the most widely used Internet innovations around the world, I'm pretty sure that their inventors did not expect that people would end up using them so differently, and sometimes even for all sorts of different purposes than the intended ones.

An example that I always like to think of is of the people in Venezuela who use Facebook Groups in a very unique and ingenious way. Venezuelans, who have long suffered from hyperinflation and a shortage of food and medical supplies, use Facebook's group feature to form what researchers have dubbed a "Solidarity Economy."[75] Instead of purchasing goods from bachaqueros—scammers who sell things at an artificially increased price—people have started to self-organize by using Facebook Groups to obtain goods at much fairer prices. Because the name of a Facebook Group can be personalized, many groups choose to use a title that includes "anti-bachaquero." This makes those groups easy to find and the closed nature of groups additionally lets people build trust once they join. In the end, Facebook Groups have allowed people to fight back against inflation, by their exchanging goods at reasonable prices in a country where the government has failed to introduce reform. In line with this, the researchers from the Georgia Institute of Technology who reported on this phenomenon concluded: "When Mark Zuckerberg first created a social site for college students at elite universities in 2004, he could not have anticipated that just over a decade later, his platform would play a significant role in basic survival for individuals in a prolonged economic crisis thousands of miles away in South America."[75, p.10]

What you see here is that people invent what I would call "coping mechanisms" to make a technology product better fit their own cultural frame. One example is simply the number of connections people form on social networking platforms. On Facebook, people from predominantly collectivistic countries such as India, Namibia, and Brazil tend to have more connections, partly because they are more likely to befriend strangers and friends of friends in addition to "actual," offline friends.[192,215] This often results in more tight-knit network structures in these countries, with most of your friends also being connected to each other. For example, when the researchers Jinkyung Na, Michal Kosinski, and David Stillwell analyzed 26,847 Facebook users across 49 countries, they found that those in individualistic countries were more likely to have ego-centric networks, in which members of the networks are connected to you but not necessarily to each other.[192] Users from collectivistic countries had more far-reaching, tight-knit networks. The authors attributed this finding to the independent nature of individualists, who prefer standing out to blending in: "When a new tool is introduced in different cultural contexts, preexisting cultural practices, rather than the tool itself, are likely to determine its usage."[192]

I'd say the reality may be a mixture of the two conclusions, because more recent work has shown the opposite on closed network platforms, where content is mostly shared with only a small number of trusted users. Snapchat is one example, an instant messaging app that lets users share messages and pictures that delete themselves after a short time. Most "snaps" are shared with another user or a group of selected friends or acquaintances. Unlike on Facebook, Snapchat users from individualistic countries, such as the US, have more far-reaching friendship networks than those in more collectivist countries, such as Japan.[247] The researchers explain this with the relatively high *relational mobility* in countries like the US—relationships with other people are determined by choice, are fairly fluid, and frequently get replaced with others.[272]

People can freely befriend new people and leave old friends behind. In contrast, Japanese society has low relational mobility. Its relationships take longer to form but are more stable over time.

Beyond adapting the size and structure of online social networks to cultural needs and values, people also find workarounds for befriending someone. It's a curiously Western idea to think that simply becoming "friends" with someone online suggests a deep and meaningful connection. And indeed, researchers have questioned whether this feels natural to interdependent people.[76] After interviewing Iraqi citizens about their use of technology in everyday life, the researchers Bryan Semaan, Bryan Dosono, and Lauren Britton discovered a deep rift between the Iraqis' desire to "save face" and uphold a positive collective identity—values that are in line with their high-context and collectivistic culture—and the design of technologies that commonly encourage serendipitous connections.[245] Taking Facebook as an example, the authors describe how the core value of the social network site is centered around openness: "Facebook facilitates the process of creating loose connections among its users." But with that, it is ill-suited for people who don't share this value. If you are from a hierarchical culture, one with a high power distance, you probably think it's a pretty awkward idea to send a friend request to just about anyone, inviting someone into a loose connection with you. Maybe you even see it as a provocation to uproot societal norms. When the researchers Anicia Peters, Heike Winschiers-Theophilus, and Brian Mennecke studied how Facebook is used in Namibia, this was one of the key insights they found: Facebook is designed for flat societies, not for Namibians and many others who usually practice a fairly high power distance with parents, village elders, and older relatives.[215] Of course, Facebook was initially designed for US students—making everyone connect by "befriending" each other made perfect sense among a student population in a country in which societal structures are relatively flat. But in Namibia, to become "friends" with your village elder is to clash with the dominant cultural norms. To compensate for Facebook's missing power distance, Namibians came up with a workaround, a coping mechanism: When they send a friendship request, they accept that they are in a more submissive role for all subsequent interactions. It's a set of unspoken rules that determine who is supposed to initiate an online connection and how they can later interact with each other. While this is still awkward, Namibians use friendship requests to reinforce societal ranks.

Digital Self-Presentation

Of course, it's not only important *whom* we connect with on social media, but also *what* we share with them. What they share reveals something very interesting about a person. Is it baby photos? Cats? Recent successes? As it turns out, there are large differences across countries and cultures in what people post online, and many of these differences show that people shape technology to fit their own cultural contexts.

Take people from Arab Muslim countries who often share Quran verses on Twitter, the social networking service that is now officially called X. It's a sort of digital religious movement. Together with other researchers, Norah Abokhodair, a senior program manager at Microsoft and a native Arab Muslim, asked herself how Twitter is being used to support religion and why.[1] What role does this online expression of religious views play in people's lives? She ended up analyzing 2.6 million tweets that contained Quran verses and conducting surveys and interviews with some of the Twitter users who had shared them. Her results are a window into a completely different world for me. They paint a picture of a transformation of Islamic rituals, which are now increasingly practiced online. For instance, almost half of the survey participants in Abokhodair's study said that they tweet Quran verses. I can only assume that before online social media existed, people weren't sending each other letters with Quran verses quite as regularly. So this sharing of verses in such a capacity is a new thing. Even local businesses share Quran verses, which Abokhodair writes might be to appeal to their customers. But the motivations seem to vary: Sometimes the verses are tweeted as an expression of solidarity and support, sometimes as good deeds, and often to remind the tweeter, or others, of the meaning of specific verses. For Arab Muslims, X has become a virtual sacramental space, which lets them engage in new forms of religious expression with people well beyond their local communities.

It's probably fair to say that there is a certain amount of digital self-presentation in this. By sharing religious verses on Twitter, people have a newfound chance to show the world (or at least the Internet) what they believe and value. In Islamic societies, the digital self-representation has become part of people's attempt to collect good deeds. I find it quite impressive how people carefully construct their social identities in this way, especially because managing how others perceive ourselves online is no easy task. What is striking in Arab Muslim cultures is that creating these social identities follows the rules of Islamic and collectivist cultures. It's not just about satisfying your own needs for self-presentation, but mostly about being responsible for how others close to you will be seen, such as your extended family, tribe, country, or religious group. When Abokhodair studied how Qatari and Saudi participants share photos on social media,[2] she found that they are indeed incredibly thoughtful about how the content they share may impact the groups that they are part of. In contrast to individualists, people who grow up in collectivist cultures are much more likely to tie their own identity to a larger unit. I would not be surprised if this form of managing social identity is stronger in Arab Muslim cultures, in which people are required to conform to Islamic rules, than in other collectivist societies.

Managing others' impressions also means using platform functionalities in ways that fit their specific cultural context. Sarah Vieweg and Adam Hodges were interested in how Qataris do this when managing their public images through social media.[287] There were several findings from their work that particularly resonated with me. One was that none of the female participants in their study posted pictures of their whole face. Maybe half their face, but never more than that. It's a very smart way of maintaining modesty, while still being able to go in on the fun part of social media (you guessed it, collecting "likes"). It is fascinating that Qatari participants in their study frequently reported having two social media accounts on any given platform: One for managing their "official" social identity that conforms with traditional Islamic values and in which they would never post photos of themselves, and one shared with their close friends, which they use to share posts and photos (showing parts of their faces and bodies) that their family or other people should not see. My kids aren't old enough for me to know whether this is something teenagers do all over the world—my hunch is many probably do. But I would also assume that *adults* in Western societies usually do not feel the need to have two accounts to promote two different social identities. As Vieweg and Hodges point out, the rules in Qatar and other Islamic societies are very different from those in other cultures. Following them not only puts an additional burden on Qataris (presumably mostly on Qatari women, who have to manage several accounts) but also means that for Arab Muslims the self-presentation on social media needs to carefully balance their own identity and the norms of their family and collective group.

Think about this for a moment. When you use social media, how much do you think about how this may reflect on your family, community, or religious group? How much do you think is your own identity tied to others? Our test on LabintheWild should give you a pretty good estimate. Try to do it as quickly as you can.

LabintheWild study #5: https://labinthewild.org/bookstudy5

Have you completed the test? You may have been surprised to be asked twenty times how you would answer the question, "Who am I?" It's a slightly altered version of a test with a very memorable yet uninventive name, the "Twenty-Statements Test," developed by Manford Kuhn and Thomas McPartland in 1954 to study self-attitudes.[149] The rationale for this test is that the more people perceive themselves as connected with others, the more their statements will refer to their roles in society, within their communities, or within their family. Those who include more of these relationships are likely to have a more collectivistic self-concept. In contrast, if someone rarely mentions their collective identity and instead primarily focuses on personal characteristics, they can be considered more individualist. Like most attempts to quantify psychological concepts, the Twenty-Statements Test has received its fair share of criticism. But it has helped researchers find interesting results that I think are worth pointing out. For example, psychologists Ma and Schoeneman used it to compare Americans to Kenyans, including members of the Maasai and Samburu tribes.[171] How do you think this comparison played out? In short, the two researchers found that the more Western someone is, the more they use individualistic statements. But let me break this down a bit. When given the Twenty-Statements Test, almost half of the statements made by American college students referred to personal characteristics and only 7% were about roles and memberships, such as references to their kinship or their occupation. That's less than one statement about roles and memberships per person, on average. Interestingly, Kenyan college students in Nairobi dedicated a similarly low number of statements, 6%, to roles and memberships, and 38% to personal characteristics. The researchers hypothesized this would happen because people living in urban centers are often more individualistic, even if the rest of their country tends to be collectivistic. In addition, they reasoned, Kenyan college students have likely been Westernized by attending schools shaped by British colonialism.

It's fascinating, then, that the results were completely flipped for members of the Maasai and Samburu tribes. Less than 2% of their statements were about personal characteristics, but an incredible 58% of statements listed by Maasai and 68% by Samburu described their roles and memberships in their communities. American and, to a slightly lesser extent, Kenyan students have an individualistic, egocentric self-concept. Maasai and Samburu think of themselves much more as part of a larger group. Their social identity is at the forefront of their mind. I'm going out on a limb here, but I could bet that the Maasai and Samburu would be much more concerned with representing their family and tribe when sharing on social media than with their own individual identity.

There is another interesting assumption we often make: that social media users themselves make decisions about what to share and with whom to share it. But what if this autonomy of sharing does not apply universally? This is what a team of researchers led by Nicola

Bidwell, at the University of Namibia, found.[40] When designing a social media platform for information sharing across low-income communities in South Africa's rural Eastern Cape province, the research team subconsciously presumed that anyone would be able to sign up for the platform and share information. After all, at the time of their work around 2010, the majority of people in Nyandeni (a local municipality in the Eastern Cape) owned mobile phones and were increasingly interested in digital technology. The researchers deployed their local social media platform in Mankosi, a rural traditional community within Nyandeni municipality that consisted of several villages with 580 households at the time. Mankosi is governed by a tribal authority, which includes a headman in each village. What Bidwell and her colleagues found is that this headman restricted who could access the tablet on which the social media platform was running and who could share. People higher up in the community hierarchy were allowed, but not those low in this hierarchy. In other words, a social media platform that assumes a flat hierarchy in society is up against local customs. Or, as the researchers put it: "Design continues a history of colonialism and embeds meanings in media that disrupt existing communication practices."[39] We will talk more about this kind of digital neocolonialism in Chapter 9.

What I take away from these examples is that social network platforms are commonly not designed to support the needs of people in certain countries and cultures. Accidentally oversharing with the broad public is all too easy, because people can't easily select who they want to share content with. For example, Facebook's audience selector is explained on hundreds of non-Facebook websites because people find it terribly confusing, if they know about it at all. If you've ever used it, you'll know that it requires a huge amount of cognitive effort to decide whom to share specific content with. And, of course, having two accounts on Facebook and on most social network platforms is against their policies. Seeing the "real you" is baked into the values and goals of any of these platforms, because what do you have to hide? In other words, social media platforms were not designed to support impression management in cultures in which traditional values, the need to represent one's own collective, and the natural desire to form an individual identity all compete with each other.

Social media also has the built-in assumption that we want to use it to show off how distinct we are from others. That it is all about presenting ourselves and projecting that our identity is independent of others'. Most platforms support posting as an individual, not as a member of a group. Likes, followers, views, retweets, and so forth reward an individual's post, not that of a larger collective. Even if people set up a group account, there is ultimately one account administrator and one login. In other words, the design is highly focused on supporting predominantly Western ideals. As we saw in this chapter, this is true not only for social media applications but also for a range of other digital technology inventions.

That's why I think research on cross-cultural differences, and generally the use of digital technology in various cultures, is so important. It tells us that people can have a wide range of goals when using technology. And while people can sometimes shape technology to suit their cultural contexts, this shaping can only go so far if the technology design cards are stacked against them.

6

Tell Me Where You Live and I'll Tell You What You Like

I already told you about some of my experiences in Rwanda earlier in this book. But what I had conveniently left out is the embarrassment I felt when I naively assumed that people in Rwanda must share my visual preferences for subdued colors and low visual complexity. "Really? Rwandans don't like it when the interface is just gray and white with black text?" I remember thinking to myself. I had just revealed a software prototype to a few people in the Rwandan agricultural ministry, and they had politely told me that it looked uninviting and dull. (In hindsight, it did look like I had forgotten to apply some colors, which I thought would make it look streamlined and efficient.) In my defense, it was rare back then to learn about visual design and people's varying visual preferences when studying computer science. In fact, it is still rare today. But it dawned on me then that what we find visually appealing, just like so many other things, is shaped by our experiences.

Ever since, I've been working on de-graying myself and the Internet. I even decided to explore the topic of visual appeal in more depth during my PhD years. In one of my projects, I asked university students in Switzerland, Rwanda, and Thailand to compose their own interface given a few options.[225] I assumed that across these countries, people's visual preferences must be quite different. Given an empty outline of a specific web application (I chose a to-do list application to avoid their having specific expectations about its look), what choices would participants make when designing it? To make their decisions and our evaluation more manageable, I gave them three choices for various design options, such as the brightness of colors, the overall colorfulness, visual complexity, the amount of support and guidance, and how the information is organized.

I still remember not knowing what to expect from this study. The participants were all students who regularly used the Internet. They grew up and lived in different countries, but would this really mean they had different preferences for user interface design? After all, most were using Microsoft Windows as their operating system, so likely they had gotten used to the primary-colored design that it featured. In short, I thought assuming they would have vastly different preferences across countries might be too far-fetched. Do you think it is?

Well, it wasn't as crazy an idea as I had thought: Within each country, participants tended to choose similar designs. But across countries, there were huge differences. You can see the gist of it in Figure 6.1. The Swiss participants composed an interface that looked a bit like the Credit Suisse web page, one of the major banking sites in their country: Mainly white with a little bit of blue and gray sprinkled in and definitely no clutter. (Prior work had found that German participants trust websites featuring the color blue more than those featuring yellow and gray,[62] and this may be similar for Swiss users.) Their choices clearly showed a no-nonsense mentality toward user interface design. Thai participants chose the brightest color scheme of all, with different parts of the interface bordered by bright pink, green, and turquoise on a light green background. They also preferred playful icons to depict different categories over plain text. Rwandans chose the highest information density, with both icons and text descriptions showing which category each to-do belongs to. They picked an earthy color scheme with green, brown, and orange tones. And unlike the Swiss and Thai groups, they chose maximum support by adding a wizard to their user interface. The wizard was a little bit like Microsoft Clippy, that little office clip from the 1990s that would constantly ask you whether you needed help writing a letter. One of the participants later told me that he thought Clippy was quite popular in Rwanda—it gave them a buddy

(a) Switzerland

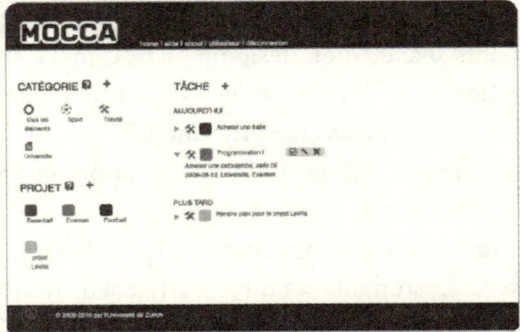

(b) Thailand

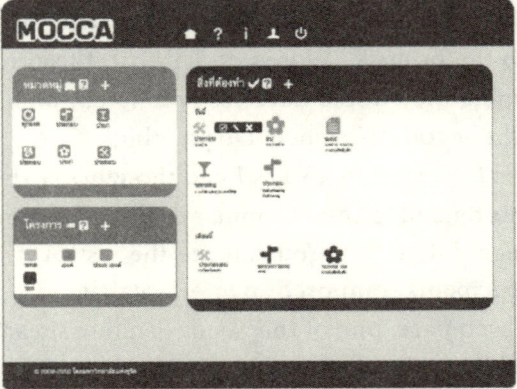

(c) Rwanda

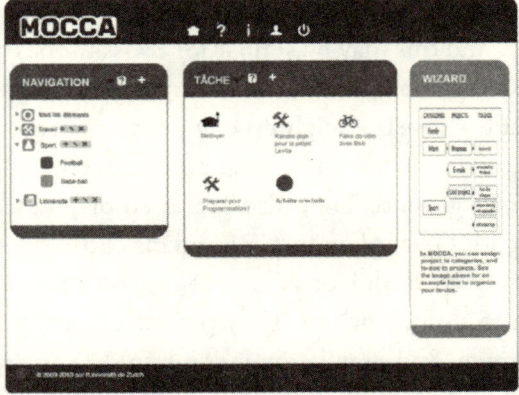

FIGURE 6.1. Participants from Switzerland, Thailand, and Rwanda chose different elements to compose a user interface in one of our studies, resulting in distinct visual designs and functionality.[225]

to connect with while working on the computer, he said. Contrast this with the polarizing and frequently hostile reactions Clippy got in many Western countries!

I think it is interesting that people from the same country seem to share similar visual preferences for user interfaces. Perhaps this is because they are used to seeing similar things in their daily lives. An acquired preference, similarly to how some of us learn to like beer or coffee. Or perhaps it is related to how we learn to process information more analytically or holistically, depending on our upbringing, as we learned in Chapter 3.

Culturally Adaptive User Interfaces

There was another exciting insight I gained from this study: Many of the choices participants made could be anticipated from a bit of knowledge about their cultural background. Prior to this study, several researchers had connected differences in websites across countries to certain attributes of culture. They found that some countries' websites have specific visual characteristics, or "cultural markers," as researchers Barber and Badre called it in their 1998 paper on *culturability*.[25] They and others linked many of these cultural markers to cultural dimensions, such as those by Hofstede. For example, Malaysian university websites were found to feature authority figures and power symbols, in line with the country's high power distance score.[103] A high power distance also seems to correlate with a more hierarchical information structure in websites.[189] Researchers found that Japanese Forbes 500 company websites commonly included images showing traditional and distinct gender roles as well as strong colors, which one could attribute to the country's high masculinity score.[253] And Chinese websites, also drawn from the Forbes 500 list, often showed more images with a family theme, in line with a collectivistic, interdependent culture.[254] Aaron Marcus, a UX pioneer who was one of the first to publish about the connection between culture and user interface design in the early 2000s, had additionally speculated that people from highly individualist societies might prefer a range of choices and functionalities, but that these should be neatly organized to reduce clutter.[176] While these design recommendations remained speculations (Marcus and other researchers compared several existing websites before making their observations), recall how the Swiss participants in our study chose a very simple interface. Could it be that there is indeed a connection between cultural dimensions and website design?

I was intrigued. If there are certain connections between cultural dimensions and design, it may be possible to provide people with designs that are more aligned with their cultural experiences. What if we used rules to predict what someone may like and redesign websites accordingly? For example, what if we gave someone from an individualistic culture an interface with a low information density, and someone from a collectivistic culture an interface that is less structured? Would these users like these interfaces better and find them more intuitive to use? Would they maybe even be better at using them compared to one that is not culturally adapted to them?

When I presented this idea to my PhD advisor, Abraham (Avi) Bernstein, I earned a skeptical, albeit encouraging, look. One of the many things Avi taught me is that it is always good to have a Plan B when doing research. What are you going to do if the link between cultural dimensions and interface design isn't as strong as prior work seemed to suggest? What if the adaptation rules you derive from it don't work out? Clearly, I needed a Plan B if I ever wanted to finish my PhD. But I was also stubborn enough that my Plan B ended up being a bit halfhearted—if the experiment didn't work, I thought, I would just have to create new adaptation rules and test it again. In other words, I had no doubt that we'd be able to predict visual preferences in some way or another, and that culture (or at least country of origin) had something to do with it.

There were three experiments we ended up doing. In the first one,[225] we generated a number of adaptation rules from previous studies on the link between culture and website design. For example, a high uncertainty avoidance score in Hofstede's cultural dimensions would trigger maximum support and guidance. A low masculinity score (which, according to Hofstede, describes the gap between men's and women's values in a society, with a low score suggesting the values are fairly equal) would trigger pastel colors with little saturation while a high masculinity score would trigger highly saturated, contrasting colors. Again, I didn't make this up. These were all design recommendations other researchers had made, usually after analyzing websites across countries. But would these recommendations generalize to other countries that had not previously been studied?

It turned out that these researchers had hit the nail on the head. The adaptation rules correctly predicted the majority of user interface choices by our Swiss and Thai participants and they were much better at doing so than randomly selecting an interface choice. Rwandans' user interface preferences were the least predictable with our rules, though Rwandans made very similar choices when selecting user interface elements. We figured this was because our adaptation rules relied on mappings between Hofstede's dimensions and user interface designs, drawn from studies that mostly focused on website comparisons between North America, Asia, and Europe. In other words, there was very little data about Rwandans' visual preferences that our adaptation rules could build on.

In a second study,[223] we were curious whether the adaptation rules would be still useful for people who had lived in several places, for culturally ambiguous users, as we fondly referred to them. We recruited our participants in Switzerland, but the vast majority had lived in at least one other country before. The tricky part was finding a way to estimate how these experiences in other countries had influenced their levels of individualism, power distance, masculinity, and so on. We decided on an admittedly very crude way of calculating this: We simply multiplied the fraction of time someone had lived in a certain country relative to their overall age with each of Hofstede's dimensions. We knew this would be a very rough approximation of someone's national culture, but we also thought that this may be better than giving people an interface optimized for Swiss users despite their having been exposed to other countries and cultures. (Notably, conventional approaches to localization would simply give people who access their websites from

Switzerland the Swiss version, since they usually rely on the user's IP address.)

Our results were surprisingly good—and definitely good enough to make me discount the need for a Plan B. The adaptation rules again correctly predicted the majority of the interface design choices our culturally ambiguous participants made. This was despite the fact that participants' choices of user interface elements were fairly evenly distributed across the three options. And even if the adaptation rules were wrong, the participants' choices were usually not hugely different. I think what I found perhaps most interesting about these results was that all of our participants had been living in Switzerland for at least nine months, yet their visual preferences were hugely different from those of the Swiss. But the longer someone had lived in Switzerland, the more they aligned with the Swiss preference for simple interfaces. This also means that just using the Hofstede dimensions assigned to their country of origin was not enough; their moving around clearly influenced their visual preferences.

All of this was encouraging enough for us to decide to do a third study, this time with a fully working web application. We now knew that we could (sort of) predict people's design choices. But would they also be better and happier when interacting with the interface they chose compared to another version?

There was already some indication that people prefer designs that are made by people from the same country and that they can use them more intuitively.[77] When researchers gave Chinese and American participants a website designed by someone from their own country and one designed by a person from another country (without telling them that this was the case), both participant groups were much faster interacting with the website designed by someone from their own country than with the foreign-made one.[78] In fact, Chinese participants found the website by a US designer less intuitive to use, and vice versa. It seems that what we like is determined by where we live, and what we can intuitively use is determined by what we like. But would we see that people are also better at interacting with websites that are adapted to fit their needs?

I jumped in by recruiting 41 international students and expats in Switzerland who had previously lived in two to five countries and represented 25 nationalities. Each one of them was asked to complete several tasks with two different versions of the web application: with the US version and with a culturally personalized version based on the adaptation rules. As we often do in these studies, the order in which participants interacted with these two versions was alternated between them to avoid giving one version an upper edge. While they worked on the tasks, we recorded how long they took, how many errors they made, and the number of clicks they made before completing a task. Participants filled out a questionnaire after each version, so we could later compare their impressions of working with the different versions.

Of course, there's no obvious reason why participants would be objectively faster when completing a task with a version that is culturally personalized to them as opposed to working with the US version. Except if the human mind subconsciously knows what's best for us. What we find most intuitive. There's actually quite a bit of evidence that what we find visually appealing affects other impressions. Researchers call this a *halo effect*. In 2000, Noam Tractinsky and his colleagues at Ben Gurion University in Israel published a highly influential paper titled "What is beautiful is usable," in which they showed that this effect is quite profound.[277] Participants interacting with an ATM that they found aesthetically appealing rated the system's usability as much higher than that of a system they found less appealing. Whether they used an ATM that was deliberately slower and had faulty buttons (as in the low usability condition) or whether one that was working smoothly did not make a difference. In one of our LabintheWild experiments, we found that visual appeal even positively influences how quickly people can find information: While people are generally faster at finding information on very simple websites than on those that are cluttered (no surprise here, since simple websites have fewer items to look through), those who usually prefer highly complex websites aren't slowed down by them as much as people who prefer simple websites.[33] So if one needs a visually complex website (such as those common for online newspapers or e-commerce stores), then people who don't find them visually appealing will be more negatively impacted than those who do like them (and who are, presumably, used to the more complex design). In other words, what is beautiful is usable, independent of a product's actual usability. Other researchers have found that appeal also affects how much we trust a site[161] or whether we intend to comply with security warning messages.[251] All of this research shows that it really is important to get the design right.

The results of our study were also likely influenced by a halo effect. Our culturally ambiguous participants found the personalized version more aesthetically appealing than the US version and this perception nicely correlated with their impressions of its usability. But in addition to finding the personalized version more usable, they were also objectively better at using it compared to the US version: They were generally faster, needed fewer clicks to complete the tasks, and made fewer errors.[224] While the tasks should have been achievable with the same speed and number of clicks in both interface versions, something about the personalized version was more intuitive to them. In fact, using the personalized version, they were 22% faster on average. A task that took roughly 5 minutes with the US version took only 4 minutes with the personalized version. Imagine how this would play out in your daily computer usage if you saved that much time!

What all of this shows us is that people's visual preferences are, to some extent, linked to their cultural background, and this makes them somewhat predictable. Local designers can often anticipate these preferences and produce designs that are intuitively usable. But following cultural adaptation rules seems to work similarly: When we personalized user interfaces to an individual's cultural background, we saw that they can work with it faster, with fewer clicks, and fewer errors. Given my German background, you can imagine that this potential gain in efficiency didn't cease to captivate me. If some websites are more usable for us than others, could there be a way to automatically foresee which one is more suitable for a given user? Or could we even somehow change designs on the fly?

Quantifying First Impressions

I was lucky to be introduced to Tom Yeh at that point. During his PhD at MIT, Tom had developed a way to analyze user interface screenshots using computer vision.[321] Could we use a similar approach to quantify elements of a website screenshot that play an outsized role in influencing visual appeal? Tom, my postdoc advisor at Harvard, Krzysztof Gajos, and I feverishly discussed how this could possibly work. We already knew from our prior work and that of others that there are a few of those features: a website's colors, its visual complexity, and whether elements are arranged symmetrically, for example. If we dissected a website screenshot to turn these features into numbers, we reasoned, it would give us a basis to ultimately predict website visual appeal. Over the next few years, we spent a lot of hours on the phone and video chatting with several other collaborators to develop computer vision algorithms that could quantify a website's aesthetics. But we also needed to know which websites people find more or less visually appealing. How could we find out what kind of website designs people like across many more countries and cultures without relying on cultural dimensions? Honestly, I really struggled with this question. Traveling to all sorts of countries to find out would be a dream, but seemed infeasible (and, well, expensive). Recruiting people on the Internet to participate in my studies was more doable, though it did mean that we first had to build LabintheWild. This is where it all began.

In the study we ended up putting on the newly launched LabintheWild, we asked people to view a set of websites and rate them on visual appeal. Our goal was to find out what predicts visual appeal. Would we see that people's website design preferences depend on where they come from? Feel free to test this out yourself.

LabintheWild study #6: https://labinthewild.org/bookstudy6

One thing almost everyone tells me after participating in this study is that the websites flashed way too fast. They certainly appear only very briefly—you only have 500 ms to view them before we ask you for your opinion on the website's visual appeal. But psychologist Gitte Lindgaard, a professor at Carleton University in Canada, found that people actually form a reliable first impression of websites within just 50 ms.[162] So, even shorter than in our study! If I asked you again about a specific website after a few minutes, you'd be very likely to have the same response. And while we all think this short amount of time is not enough to take in a website, that's actually the whole point of it. Your first impression is what counts. Reading a website's content can of course change your opinion of a website later on, but it is this initial split second in which we make up our minds whether we are going to trust its information. And whether we think it is a usable site. Or whether we are going to leave. (We will discuss in Chapter 7 that the way technology, such as a website, communicates with you in a culturally sensitive way certainly matters, too.)

Back to the study. You just saw many different websites that were a subset of a much larger number of websites that we had collected to study visual appeal. We excluded highly popular websites that many people might recognize in order to capture people's true first impressions of a design. Now, given all participants' ratings of visual appeal, we could tell you which website people seem to like most. That's certainly interesting, especially if I told you that across all participants, the website shown in Figure 6.2 was rated highest. But remember that one of our goals was to find out what exactly it is about a website's design that people may like most or least. Our way to achieve this was to use computer vision algorithms that helped us quantify a website's design.[229] We basically reduced a website's design to numbers. For example, our algorithms count the number of pixels that contain a certain color and the number of text and image areas, and they assess how saturated the colors are. They also subdivide each website into blocks and keep subdividing each block until a block no longer contains pixels of a different color. A large block means that the website has a large region that is homogeneously colored, as when much of the website has a white

FIGURE 6.2. The website screenshot that received the highest average visual appeal rating in our study.[226]

background. Many small blocks mean that the website is kind of busy, or visually complex. Perhaps there are lots of content regions, different colors, or images. Depending on how these blocks are distributed on the site, we can also tell where the website is most "cluttered" or whether the website is symmetrical.

After we had found a way to quantify website aesthetics, we set out to compare visual appeal across countries. At this point, our LabintheWild study had collected data from around 40K participants.[226] Together with Krzysztof Gajos at Harvard, I built a model that could tell us which design factors and which demographic variables play a role in people's visual appeal. Tell us where you are from—and give us information about your age, education level, and gender—and we will tell you what you like.

What do you think we found? Do you think people's visual preferences varied a lot or not so much? Let's start with something that seems to be a shared preference across the world: People react most strongly to the visual complexity of a site, and only then to the overall colorfulness, specific colors, and other design choices. People don't like extremely simple websites, nor do they like those that are highly complex. Intuitively, we seem to know that clutter can be bad for finding information, but so is not seeing much at all. Scientists call this an inverted U shape, which basically means that there is only a narrow range of visual complexity at the tip of the upended U where visual appeal peaks. Our data also showed an inverted U shape for colorfulness, where websites that have an extremely low or extremely high colorfulness are usually disliked.

That's about where our shared visual preferences end. Beyond this, I cannot tell you how to adjust your website so that it is universally liked, because it depends. In fact, we found that a person's age, gender, education level, and yes, country, are highly predictive of website visual appeal. Have a look at Figure 6.3, which shows a ranking of the 43 countries for which we had sufficient data to do meaningful statistical comparisons. You'll see that Russians and Finns prefer the lowest visual complexity and colorfulness in websites: The websites they rated most highly had a visual complexity of 2.5 and an overall colorfulness of 3.6, both on a scale of 1–9. You can see their most liked websites in Figure 6.4. In contrast, there are countries where the peak appeal was only achieved when the website's score for complexity and colorfulness doubled, for example, for Serbia or Macedonia whose most-liked website designs, also shown in Figure 6.4, are quite a bit busier. Put differently, visual preferences vary a lot!

Let's think about this for a moment. If we designed a website to be visually appealing for Russians, we should minimize text and the number of images and colors. A white background seems to appeal to Russians the most. But, according to our data, that same website would be perceived as way too minimalist by the majority of people across the world. The average person in Bosnia-Herzegovina and Serbia, or in Chile and Mexico, for instance, would likely get a fairly bad first impression when seeing it. And as we know from prior work,[161] this may lead to their finding the site untrustworthy and unusable, or they may leave the website to begin with.

When we looked at the results more closely, we uncovered something that I found endlessly fascinating: Visual preferences for websites were similar in neighboring countries or in countries that have similar roots. You might have already suspected this when looking at Figure 6.4,

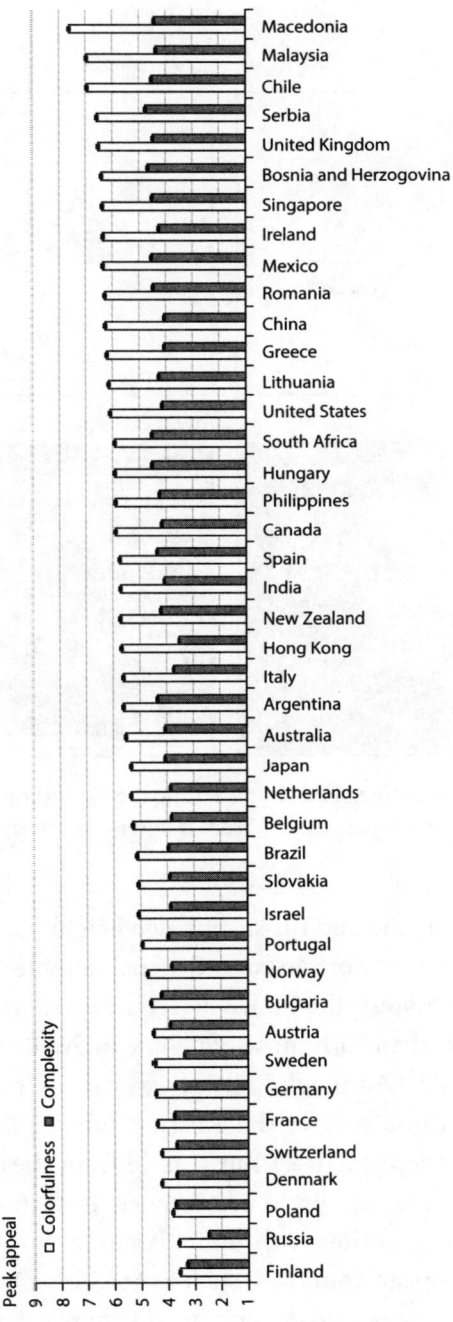

FIGURE 6.3. Colorfulness and visual complexity scores of peak appeal for different countries. Figure from Reinecke and Gajos.[226]

(a) Finland

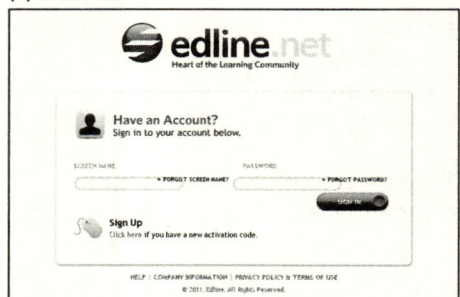

(b) Russia

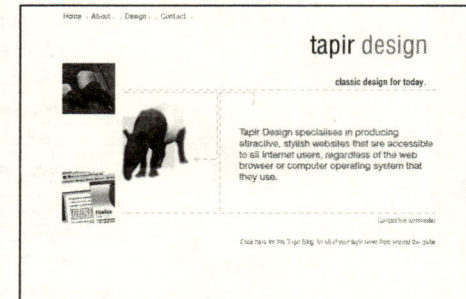

(c) Serbia

(d) Macedonia

FIGURE 6.4. Website screenshots rated among the highest on visual aesthetics at first sight by our study participants from different countries.

which includes Finland and Russia, the two neighboring countries that preferred the lowest colorfulness and visual complexity. And at least in terms of colorfulness, the Finns' preferences are more aligned with those of Russians than with those of people in Sweden, which borders on its western side. (Although Finland was part of Sweden for almost 500 years, it became part of the Russian Empire from 1809–1917 before gaining independence.) But people from these two countries are not alone in sharing visual preferences. In fact, our data clearly shows clusters of countries with similar website design tastes. People in Northern European countries, such as Sweden, Denmark, Switzerland, France, Germany, and Austria, tend to prefer fairly simple websites. People in Southern European countries, including Italy, Spain, Greece, and Romania, prefer them more colorful. Countries that used to be republics of the former Yugoslavia, like Bosnia and Herzegovina, Serbia, and Macedonia, all ranked high in a preference for complex and colorful websites. And core countries of the Anglosphere (Australia, Canada, New Zealand, and the US) also shared similar visual preferences—though, interestingly, people in the UK found highly complex and colorful websites more visually appealing than the rest of the Anglo-Saxon countries.

One of the exciting outcomes of this work is that we can now take any website screenshot and predict whether someone will like it or not. The result is not going to be perfect, but if it helps us avoid the most grueling websites, I think that's a win.

Biases in Website Designs

There is another implication of these results. Given that so many people in this world prefer fairly simple, but colorful, websites, many sites are not visually appealing to them. In fact, several of the most popular sites, from Google.com to Wikipedia.org, are too plain and either too simple or too text-heavy for the majority of participants in our study. This means that some people for whom the design is less visually appealing may be actively disadvantaged. Maybe it doesn't matter so much if it is just one unappealing site. But most of us interact with more than 100 websites per day. (My smartphone recently told me I browse more than I sleep!) Cumulatively, any kind of negative impression adds up.

Now, of course, we all often accept using websites that may not be optimally visually appealing to us. Google's search engine is perhaps one of those. Many of us use it not because we love seeing a list of blue-colored links that is rather lacking in creative design but because it gets us the results we want. (Well, sometimes.) You may have a choice here, but you end up using certain websites because they offer a unique service, they have a high market share so you assume they must be good, or maybe there is a network effect—all your friends use them, so there's no point in going elsewhere. But there are a surprisingly large number of websites where you do have a choice. Looking for information about your next travel destination? About the latest sports news? There are a gazillion websites out there that offer similar content. I constantly find myself quickly closing a browser tab when encountering new websites that look completely untrustworthy to me. (This happened to me more when I had newly arrived in the US than now . . . I clearly got used to some of the more questionable design choices here!) I also often find myself looking for information on a website but feeling like it's not designed for me.

And it seems I'm not alone: In 2018, Danaë Metaxa-Kakavouli and a team of researchers from Stanford University and Brown University published a paper in which they demonstrated that the visual design of web pages for an introductory computer science course can negatively affect whether people feel they would belong in the course, are interested in the course, or are interested in computer science and programming.[185] They randomly assigned participants from the US to one of two website designs: One had a nature-themed interface with a gender-neutral design, and the other one conveyed a nerdy image of computer science, with a large *Star Trek* background image and a black background. Apart from this aesthetic difference, the websites had completely identical content. And guess what? Compared to male participants, females in the study reported feeling much less like they would belong in the course represented by the latter design, and being much less interested in studying computer science more broadly. There was no difference in the responses of male and female participants for the gender-neutral nature-themed design.

Let me be clear about the significance of the finding. I think the implications go well beyond a reduced intent to sign up for a computer science class. As Metaxa-Kakavouli and her colleagues stated in their paper: "Biases that reduce a user's sense of belonging could unconsciously discourage her from taking STEM courses, applying for a job, or voicing her opinion online." Obviously, it shouldn't be in anyone's interest to disadvantage parts of the population, but this requires making careful decisions to reduce bias.

This brings me back to the lack of data that has plagued cross-cultural research. There's just not much systematic knowledge out there about what different people within and across countries like. For example, we don't yet know whether more subtle design changes than adding a *Star Trek* image, like adding a bit of color here and there, would have similarly serious consequences. Designers are also pretty much on their own when it comes to anticipating how people will interpret colors. (As my PhD student Rock Yuren Pang told me, people in China would perceive their name written in red as a death threat . . . better watch out for that next time you implement an error message!) But based on all the studies I have presented to you in this chapter, I think we know enough to at least be aware that visual design choices can affect people differently. Some will like a website and be just fine using it while others will not find it visually appealing, with downstream effects on their trust, engagement, feeling of belonging, compliance with warning messages, and so on. To fix this, it's not going to be enough to run A/B tests until we've found the optimal location or color of a button. Instead, we need to rethink our design choices more holistically. This could include distributing the content to several subpages to lower a website's visual complexity, moving away from templated websites that follow a specific branding to increase colorfulness, or changing images to be more culturally meaningful and appropriate. And, most importantly, we need to be aware of the potential impact of our design choices on people around the world.

7

Communicating with and through Technology

It's sometimes hard to believe how quickly technology changes our habits and norms. When the instant messaging and online collaboration program Slack became available in 2013, many people in organizations and businesses, myself included, pretty soon started using it on a daily basis. Its goals must have resonated with many: Improving productivity? Check. Streamlining communications among your team? Check. Integration with email, calendar programs, and file storage? Also check. I think people were so quick to adopt Slack because they were simply tired of emails. Unlike Slack, email programs and services were late in allowing silencing messages during focused work times or nonwork hours. And, with email technically being the first descendant of the letter, there was always a certain formality required when writing one. To this day, I find it odd to write an email without first addressing the recipient, wishing them a good day, and signing off at the end. Other people will likely agree: The Japanese, for example, commonly refer to the weather and seasons in their greetings, including in emails. "At the time of fresh green foliage, I hope that this letter finds [name] both healthy and prosperous," they may write.[304] Or, "We have come to the season of autumn leaves, how do you fare?" Ultimately, this is beautifully polite and formal, yet could be inefficient given the number of emails many of us write each day.

One of the clever ideas behind Slack was that it removed these inef-ficiencies. Once people are in your workspace—Slack's name for a community or team of users—you can drop all of these formalities. In fact, the sheer design of Slack prevents you from being overly for-mal: Instead of having a subject line, From and To headers, and a large, empty textbox to type in your email (which really is all too similar to sending a letter), Slack focuses on quick chat messages that can be sent to individual team members or the larger group. Its small input field encourages efficient, to-the-point messages, not writing a whole mono-logue. My Slack workspaces are full of variants of "Joining meeting virtually today!" "Quick to-do for you: . . ." or "Hey, can we meet ear-lier today?" and literally all messages will be decorated with at least one emoji. Punctuation marks are increasingly uncommon. Saying "thank you" is often replaced with some kind of celebratory emoji gif. And, of course, all of this depends on your familiarity with others in your workspace, though I think it is fair to say that no matter what, Slack has had a huge impact on communication norms in the workplace.

To some extent, Slack has improved communication for me at work. But I've also increasingly started to see it as a minefield for commu-nication in cross-cultural teams. To fully understand what I mean, I'd like you to participate in the following LabintheWild study, which will show you how you perceive other people's messages and how they may interpret yours:

LabintheWild study #7: https://labinthewild.org/bookstudy7

Did you notice your reaction to some of the messages? Perhaps some of them even made you cringe in disbelief at how anyone could communicate this way. It's a natural response if a message does not cor-respond to what you are used to. And just like with so many things in this book, what you are used to and what surprises you will depend on how you've been socialized, on your culture. As cultural anthro-pologist Edward Hall theorized, culture influences how much context people provide when they communicate with each other, how much they read between the lines, how much they use language formality to

express values of (in)equality, or how they interpret silence and gestures.[113] While communication behavior varies depending on a given
situation, he suggested that there are essentially two types of communication styles: Low-context communication is more explicit and
direct, whereas high-context communication is implicit and indirect,
with most information being transmitted via the situational context or
other signals expressed through a person's body language. Hall placed
Swiss and Germans toward the low-context end of the scale. Direct
communication is much more common in Switzerland and Germany
than it is in, say, the US and Canada (which are still considered low
context, but communication is usually not quite as direct as with the
Swiss and Germans). Somewhere in the middle of the scale is Italy,
though it is mostly considered to have a high-context culture. Did you
know that Italian children as young as two years old already gesture
more than their Canadian counterparts?[177] Despite being affectionally
stereotyped for their tendency to talk a lot, Italians actually place much
of the meaning of a message into hand gestures and facial expressions.
It's a pretty effective way of conveying emotions, if you ask me. High-
context cultures like those of China and other Asian countries would
perhaps find Italians' gesturing a bit too dramatic. This is because they
have different strategies for adding context, such as using silence to
express disagreement or even annoyance.[112] Saying "no," for example,
is pretty rare since it sounds too aggressive or it will feel like a strong
rejection. Instead, people may resort to saying "yes" or "maybe"—or
they may simply remain silent. You can probably imagine that low-
context communicators are prone to misinterpreting these silences and
other subtleties in tone that are used to convey meaning. Similarly, high-
context communicators can perceive the directness in low-context cultures as way too aggressive. Cross-cultural communication truly is a
minefield, albeit a very fascinating one.

Beyond Hall's distinction between low and high cultures, other factors also play a role in our communication. For example, people in
societies with a high power distance communicate more formally and
politely, especially when talking to those with higher status, such as
a boss or elderly people. The languages spoken in these high power
distance societies often have complicated systems of honorifics, way

beyond the formal "you" that one would use to address someone depending on the context in German, French, or Spanish. In my experience, transitioning from the formal "you" to the normal "you" is quite a relationship milestone—in German culture, it's something that the person who is older or who has higher social status should offer, which makes for incredibly awkward moments if you don't have precise information. Likewise, you really wouldn't want to start off with the informal pronoun "du" (the informal "you") to address someone—whether you are a person or a machine—unless you are absolutely sure the other person is okay with it. Fortunately, most high power distance societies have various intricate ways of indicating and detecting power and status. Perhaps someone dresses in a specific way, introduces themselves with their last name, or sits at the head of the table. Such cultural norms work to maintain a hierarchy in our offline lives, but they are completely missing when we go online. When I discussed this with Pamela Hinds, a professor of management sciences and engineering at Stanford University and a leading expert in global teams, she remarked that the tools that global teams use to communicate with each other lack this kind of signal. They don't have a way to indicate status and power. "A lot of these technologies that we develop, like Slack, are designed to eliminate hierarchy. It really doesn't support what would be appropriate communication behavior, say, in Japan." Maybe people don't mind, and maybe they adapt to it. But the issue is that we don't know.

Communicating across Cultures

What happens if teams communicate via Slack or other instant messaging platforms? I find this to be a fascinating question. Do people experience "cultural surprises," as Judy and Gary Olson described people's reactions when they realize other people do things differently, and culture suddenly becomes visible?[208] Does screen-based messaging even out any cultural differences in communication? And do high-context communicators find other ways to convey meaning when gestures, facial expressions, silences, and, well, context, aren't as easy to include as they are in face-to-face communication?

Susan Fussell, a psychologist at Cornell University, has focused on these questions with her work on computer-mediated communication in cross-cultural teams. In one of her prior projects, she and her co-authors wondered how people from different cultural backgrounds perceive online communication. They started by asking: "What features in common instant messaging applications do people from low- and high-context cultures appreciate most?"[137] The survey they used to answer this question included 78 participants from North America, India, and East Asia (mostly Singapore), all of whom regularly used various instant messengers, from ICQ to AIM, MSN, and Yahoo. (If you have not heard of any or most of them, it is because their study was done in 2006—but these messengers were surprisingly similar to modern-day WhatsApp, or to the basic instant messaging feature in Slack, for that matter.) Independently of country, all of these participants said they send a similar number of messages per day and had similar numbers of "buddies." But the differences across country groups were compelling. When asked about the importance of having features like audio chat, video chat, and emoticons (such as the smiley face containing characters from the keyboard, :-), or ^_^ as it is commonly depicted in Japan), the North American participants generally indicated that these are not at all or only slightly important. Again, keep in mind that this was in 2006 and much has changed since then. But already then, the Indian and East Asian groups both thought that emoticons and audio/video chat were much more important than North Americans thought they were. The research team reasoned these differences may be because Indian and Asian participants naturally rely more on audio, video, and emoticons, all of which add more context than just text. In their discussion, the researchers acknowledged that the results were somewhat inconclusive; after all, it could be the non-Western scripts used in many languages in India (e.g., Hindi) and Singapore (e.g., Singaporean Mandarin, Malay, and Tamil) that contributed to the preference for using non-textual communication. But, consistently with their results, other researchers later found that Indians use many more emoticons in Internet forums when writing in English than do Germans.[216] Altogether, it seems that high-context cultures tend to use emoticons more than low-context cultures. They fill a gap in text-based online communication that would usually lack any signs of context and emotion.

It is probably fair to say that since these early years, emoticon use—and the use of its more sophisticated-looking successor, the emoji—has exploded around the world. Emojis were first invented in 1999 by the Japanese artist Shigetaka Kurita to respond to the Japanese people's desire to include images in their text messages. When Apple integrated an emoji keyboard into its operating system in 2011, emojis fully took off around the world. Since then, they have become available in pretty much any instant messaging app and have evolved to represent an overwhelmingly large number of emotions, occupations, weather conditions, traffic symbols, and so on. Their impact on our communication and on the "art of our time" has been so profound that emojis are on display at New York's Museum of Modern Art.[151] Today, people use emojis constantly and for various reasons: For most, they are a way to adjust the tone of messages; to form relationships by being used in unique ways that are understood only between two individuals or within a group; or to express identity and feelings.[144]

WeChat, the popular Chinese superapp that was launched in 2011 and is now used by more than a billion users, has added to the popular rise of emojis. Many Chinese use emojis instead of text or to subtly convey the meaning of a message.[325] In a study led by Yuan Wang at Tianjin University of Technology in China, the research team even found that people's choice of emojis depends on their audience.[296] For example, when communicating with people in higher social positions, such as teachers, parents, or managers, they may use only the most common emojis and create an image of themselves as obedient and respectful. In a country where the dominant culture is that of high power distance, Wang and her colleagues concluded that many people try to maintain and respect hierarchical social structures while carefully adding emojis that are needed to avoid misinterpretations of their messages.

Of course, the nice thing about emojis is that many of them are easy to interpret. Almost anyone will recognize the "face with tears

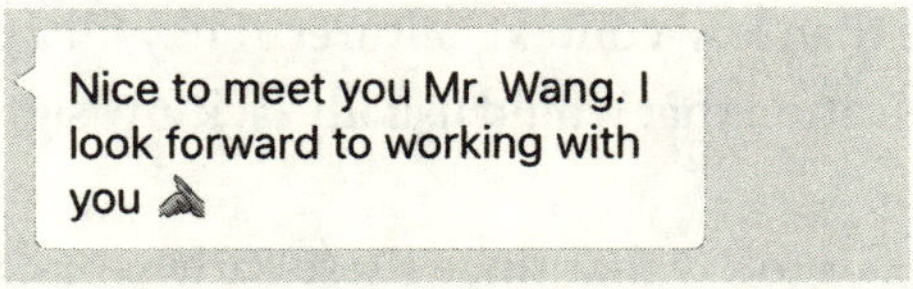

> Nice to meet you Mr. Wang. I look forward to working with you 🤜

FIGURE 7.1. A WeChat message using a "pre-fight" emoji.

of joy" 😂, which, as it turns out, is the most popular emoji in many countries, used when one finds something hilarious or maybe if one wants to make fun of oneself. (Only the French prefer sending hearts over crying with laughter.[168]) But then there are some emojis that are culture-specific and that not everyone will immediately recognize. They are primarily relevant and popular among specific groups of people, where they represent foods, artifacts, holidays, or anything else that is part of their culture. I love this about emojis, because there's always something new to discover and that can be an incentive to learn about a culture. For example, if you've ever seen a red envelope emoji in your messaging app of choice, it's a symbol for a monetary gift that the Chinese commonly give on holidays and other important festivities. Sending a red envelope emoji is a cultural gesture to symbolize luck and prosperity.[325] If you are a WeChat user, you may also regularly send and receive a "pre-fight" emoji, perhaps along with a message similar to the one shown in Figure 7.1.[324] A Westerner might think you're trying to pick a fight when you send this, but it actually represents the salute that Chinese kung fu fighters show to each other to signal respect for the opponent's skills. So it is really just meant to be a greeting, especially when meeting someone who is higher up in the social or work hierarchy.

Alas, culturally specific emojis are somewhat unevenly distributed across the world. Designer Philippe Kimura-Thollander and human-computer interaction researcher Neha Kumar from the Georgia Institute of Technology found that most of them represent Japanese and US culture, such as the Japanese Goblin, 👺, a mythical creature, or Narutomaki, 🍥, a Japanese fish cake. Other countries trail far behind and often don't have any of "their own" emojis. Maybe this wouldn't be a problem, except that people from different countries and cultures often can't correctly identify culturally specific emojis. Kimura-Thollander and Kumar think this can be a problem: "The consequence of this is that cultures outside of Japan and the West are less able to convey their daily lives through emojis and cannot participate as well in this new universal language."[144, p.12] It's a form of cultural bias—an emoji hegemony if you want—that could be easily removed if every culture had its own set of emojis.

Genmoji, the AI-generated, personalized emoji that Apple presented in June 2024, may, in the future, homogenize the playing field.[17] But until then, here's an interesting tidbit you may not have known: Anyone can suggest a new emoji to the Unicode Consortium. (Yes, the set of emojis is not dictated by Apple and Google as most people seem to assume,[144] or at least not directly.) So if you think you are missing a specific emoji in your life, please go ahead and create it. There's just one problem: Submitting new emojis to the Unicode Consortium is much less straightforward than would be desirable. In their paper, Kimura-Thollander and Kumar describe how it took them up to a day to put all the materials together for submitting just one emoji. And when they finally had everything in one place, they were at the whims of the Unicode Consortium decision makers who are heavily skewed toward Western views. (At the time of this writing, 9 of the 14 Unicode Consortium's voting members were US technology companies.[127]) Who gets a seat at the decision-making table is far from being fairly determined, and the decision-making process is far from transparent. "I recall when the bubble tea discussion was ongoing most of the senior committee members had never even had the drink and didn't understand why it should be an emoji," Kimura-Thollander, who attended a few of the emoji subcommittee meetings, told me.[143] The bubble tea emoji that Kumar and he had submitted to the Unicode Consortium has since been added, along with other emojis suggested in the paper, such as one for a sari, a women's garment worn across the Indian subcontinent, and one depicting the phoenix, an immortal mythological bird that plays important roles in ancient Greek, ancient Egyptian, and other cultures. (The phoenix symbolizes rebirth, immortality, and renewal, so it'll likely be a very popular emoji in some cultures while others may rarely use it.)

If you recall what we discussed above about the importance of emojis for expressing one's cultural identity, for bonding, and for adding nuanced meaning to a message, it's great to see emojis diversifying. Given their popularity, more culture-specific emojis are an important equalizer to Western dominance in tech. But, of course, emojis are no magic bullet to online communication, and especially not to intercultural online communication. In 2017, a team of researchers from Cornell University, led by Ge Gao, explored how people interpret the emotional meaning of each other's messages when communicating via instant messaging with an optional choice of emojis.[95] Could someone correctly identify emotions from just looking at another person's instant message? Psychologists call this ability to correctly interpret someone else's emotions *affective grounding*. In the study, the researchers randomly assigned Chinese and American students into pairs, forming ten pairs of Chinese participants, ten pairs of American participants, and ten intercultural pairs in which an American and a Chinese participant worked together. Each pair was asked to discuss an ethical dilemma, dubbed the *lifeboat task*. It's basically like the sinking of the *Titanic* on a smaller scale: Imagine nine people on a yacht that just had an accident. People will need to get on the one lifeboat there is, but (of course, there's a "but"!) it can only fit five of the nine people. The participants were asked to look at descriptions of the passengers and rank them. They then shared and discussed this ranking with their partners over instant messaging, with the goal of creating a joint ranked list that both would be happy with.

The study gave several intriguing insights. First and foremost, intercultural teams had more problems finding affective grounding than teams where both participants shared a national culture. Chinese participants, in particular, were much better at correctly interpreting whether a message sent by their partner signaled friendliness or frustration if their partner was also Chinese, as opposed to American. The Cornell research team observed that this may be because Chinese participants often adjusted their messages using emoticons and emojis—they were using them to more clearly convey what they were feeling. They also used punctuation marks, such as "...," that can indicate one is not entirely sure. Instead of directly saying "no," or "I disagree," which they perceived could lead to their losing face with their partner,[18] the Chinese were trying to more subtly indicate their opinions. In other words, they wrote in ways that are consistent with a high-context communication style. Because the Americans in the study wrote more matter-of-factly, Chinese participants often perceived the messages as signaling frustration with them, when the messages were actually just meant to directly communicate what the American participants thought. A clear signal that communication styles can clash and that the ability to add emoticons, emojis, and other elements providing context is helpful for some.

There was another result that stood out to me: Americans were not great at correctly interpreting the affective meaning of a message sent by a Chinese partner—but they were equally bad at interpreting messages by other Americans. In other words, Chinese participants were quite a bit better at finding affective grounding than Americans. What could explain this finding? Earlier research has already shown that a sample of American participants could identify emotions in online messages, so a lack of ability is unlikely the reason. Instead, based on their observations and interviews with participants, the Cornell research team speculated that Americans were less invested in understanding the emotions of their partners. They simply didn't *try*. For them, getting the task done took priority—and emojis may even make the process less efficient. In contrast, Chinese participants were very aware of the importance of building a relationship with their partners. As one of the Chinese participants explained retrospectively: "I wanted to make the partner feel we are cooperative."

Feedback to Smooth Intercultural Communication

These findings obviously matter to anyone who uses instant messengers to communicate with others, and especially to those of us who use them with people from other countries and cultures. Researchers in business communication, management, and psychology have scratched their heads for decades on how to support communication in cross-cultural teams—and not only because any communication disaster can cost companies huge amounts of money. They can also increase misunderstandings and stereotypes, and decrease trust.[9] It's an immense issue in teams that consist of people with diverse cultural backgrounds, especially if they are globally distributed.[123]

How can we mitigate friction in intercultural teams? And to what extent should we do so given that we can learn from this friction? Human-computer interaction researcher Helen He devoted her entire PhD thesis to exploring these questions. In one of her studies, she and her co-authors wondered whether it could help to make people aware of intercultural differences in communication with a team member from another culture.[119] She asked Japanese and Canadian participants to decide how to allocate $1.8 million to social causes, and to negotiate this over email in pairs of two (one Japanese and one Canadian participant). Out of five social programs, each pair was asked to choose two programs, of which the first one would get the most money. To make the task even more challenging, some of the social programs were more relevant to Canadians (those geared toward the integration of immigrants and refugees and toward the improvement of rehabilitation programs for drug addiction), while others addressed pertinent issues in Japan (regulations about workplace overtime and elder care using robots). How would Japanese and Canadians negotiate how to distribute the money over email?

Prior work had already found that the two cultures differ greatly in their negotiation style. Whereas the Japanese tend to avoid direct conflicts and instead use the face-saving strategies of indirectly suggesting any disagreements, Canadians prefer assertive negotiations and discussions of facts.[104,112] The key to making these intercultural negotiations work, He and her team thought, is to provide the participants with feedback on the language they used in their emails to their study partners. They designed the feedback in the form of graphs that would tell each participant things like, "In comparison to your study partner, how much did your email focus on building a relationship with the other person versus being task-focused?," "How much did it express your emotions, such that your team can achieve affective grounding?," and "What was the level of formality you used in your emails?" Participants were shown this feedback—a simple bar chart comparing their own email performance to that of their study partner—after each of them had sent two emails.

Now, imagine you just emailed your co-worker, and a graph popped up telling you that, well, you could do better in relationship building. I'm guessing you'd at least slightly adjust your behavior in following emails. This is exactly what happened in the study. For one, people suddenly noticed that they were communicating with someone from a different culture. The graphs made cultural differences more salient. Being presented with this feedback, Canadians also became more motivated to understand why their partner chose specific social programs and not others, and they ended up being more accommodating in the ongoing negotiation. In the end, Canadian participants yielded to their Japanese partners' choices more than when not being given feedback. The negotiations between the Japanese and Canadian participants became more equal.

It's quite a fascinating result if you think about it. What their study shows is that, all too often, people don't think about cultural differences in communication. This can lead to misunderstandings and difficulties in communicating and negotiating, and ultimately make people dissatisfied with the outcome. The good news is that highlighting these cultural differences with simple, personalized bar charts can be enough to make people aware of them and nudge them to adjust their own behavior. In other words, people don't want to be culturally insensitive; they just need a little help.

I'm pretty sure such a feedback feature could also improve intercultural communication on Slack and other online channels, whether it is for bidirectional communication or when discussing things in a larger group. If you remember what I wrote about online communities in Chapter 5, you may have already suspected that users of Stack Exchange were experiencing a variety of issues related to intercultural communication. Other researchers have found similar problems on Wikipedia, the nonprofit bringing us the world's largest and free online encyclopedia. Its community is built on an open discussion of article content via its Wikipedia Talk pages. While contributions to the discussions are supposed to "be polite" and "be positive,"[308] there is great variation in how Wikipedia editors comment on an article or suggest improvements. It's not a big surprise, given that the English Wikipedia has the largest number of editors among all Wikipedia language editions, and that they are immensely culturally diverse. But despite this diversity, comments on the English Wikipedia Talk pages are generally written as if coming straight from a low-context playbook: To the point and very much focused on the task. Indeed, a study led by Noriko Hara from Indiana University found an intriguing difference between the posts on the English Talk pages and those in other languages.[115] After analyzing 1,253 posts on Wikipedia Talk pages in four different languages, Hara and her co-authors discovered that Japanese and Malay Wikipedia Talk pages included more than four times as many polite messages as the English and Hebrew versions. In contrast, around 36% of the posts on the English and Hebrew Talk pages were disagreements, such as about facts, or the style and format of a Wikipedia article. On Japanese and Malay Talk pages, the percentage was only 6.5%. What the team of researchers found is that Japanese and Malay speakers more often write messages that include greetings, apologies, or expressions of appreciation. They generally try to keep conflicts to a minimum. Collectivistic societies really do seem to emphasize relationships over tasks.[282] For Wikipedia, making editors aware of these differences by providing feedback on cultural communication styles could be key to keeping intercultural conflicts to a minimum, especially in more heterogeneous language communities, such as that of the English Wikipedia. Rather than being taken aback by a post, people may better understand the rationale behind it and be more willing to engage in discussions. If Wikipedia wants more editors, I think this is worth a try.

Communicating with Robots

So what about when we communicate with technology or technology communicates with us? Do these cultural norms in interpersonal communication still apply? According to Malte Jung, one of the authors in the intercultural team's study and an expert in human-robot interaction, affective grounding is a concept that robot developers should be aware of because it may help us better understand how people interpret a robot's behavior.[135] Do they think what the robot says indicates that it is sad or happy? When do people think a robot sounds friendly, and how does this affect their interaction with it?

It turns out that people are indeed quite influenced by what a robot says, and how it says it—whether it speaks in a more direct, low-context manner, or more indirectly. Researchers even showed that this can affect how much people listen to it.[294] To study this, they recruited 160 students from Tsinghua University and another 160 students from Stanford University, and randomly assigned them to 80 Chinese student teams and 80 US student teams. Each team of two participants was tasked with working with a robot on setting up a chicken coop. Of all things! You won't believe how many choices one needs to make when setting up a chicken coop. Choosing the plot size, how many chickens, the chicken breed, soil type, etc. A robot who can give advice could certainly come in quite handy. (Though I'm sure there's an app for that?) Anyway, the interesting part here is not the chickens, but that the robot talked to half of the teams in an implicit, indirect way when giving recommendations. For example, it would say things like, "A bigger area allows some grass in a chicken's diet, which can make chickens healthier and increase egg production." To the other half of the teams, it instead communicated more directly, giving concrete recommendations that leave little uncertainty about the robot's opinion: "I think we should choose 75 square meters because having some grass in a chicken's diet makes chickens healthier and increases egg production." Guess how people reacted? They followed the robot's advice, but only sometimes. In fact, the US teams were more swayed by the robot when it provided direct recommendations than when it used a more implicit style of communication. And it was the opposite for Chinese teams: Chinese participants adhered to the robot's recommendations more often when it used implicit recommendations, a finding that the research team attributed to trust. Chinese participants simply trusted the robot more if it talked in an indirect style, consistent with their own high-context culture. A robot that imposed its opinions on them too directly, giving them little opportunity to collaborate on the solution, was not perceived as particularly trustworthy. The takeaway is simple: "When robots behave in more culturally normative ways, subjects are more likely to heed their recommendations," the lead author, Lin Wang, and her colleagues wrote.[294] In other words, culture matters in the design of these things.

In the chicken coop study, the robot was a stationary prototype that was able to listen and respond to basic questions. But verbal communication is only one aspect of successful cross-cultural interactions. If robots are to be accepted by humans, they also have to be adept at adjusting to cultural differences in many other customs, such as gestures, emotional expressions, or interpersonal distance behavior (the amount of space people leave between themselves and others, or between themselves and a robot, for that matter). In fact, research has shown that keeping the appropriate distance from someone else is another one of those cultural minefields that is easy to get wrong. Edward Hall called it *proxemics* when referring to people's use of space.[111] Americans' personal space, Hall suggested, is roughly 18 inches (45 cm) when standing next to someone. If someone moves inside that circle, for instance, when having a conversation, Americans may experience this as an awkward invasion of their personal space. People may also interpret other social traits from this, such as dominating behavior and status or feeling close to someone else. But this protective bubble has different sizes, depending on culture. Arabs, for example, commonly stand closer to each other than Western Europeans or Americans[303]—their personal space bubble is smaller. When asked to have a five-minute conversation with someone else from the same country, researchers found that Japanese students sat on average 40.2 inches (102 cm) apart from the other person, American students 35.4 inches (90 cm), and Venezuelans only 32.2 inches (82 cm).[262] While context matters (for example, Japanese are famously known to tightly pack into a subway), these differences are often explained by subdividing the world into *contact* and *noncontact* cultures.[111] In contact cultures, such as in many Arab, Latin American, and Southern European societies, hugging, touching, and close proximity is part of everyday life and an important signal for building and maintaining relationships. (Curiously, people tend to prefer standing closer even to strangers the higher the temperature is in the place they live.[255]) But in noncontact cultures, such as North American or Northern European countries, close contact with others outside of a few occasions, such as greetings and goodbyes, can cause considerable anxiety in people.[300]

I cannot tell you how many times I have observed cultural differences in proxemics to be an issue between people from different countries. Language barriers often play a role, but my experience has been that differences in how much contact is appropriate and wanted can have more severe consequences. In contact cultures, hugging a colleague or holding their hand a tad longer than strictly necessary, maybe even placing the other hand on their shoulder, can be perceived as a sign of affection, bonding, encouragement, and trust. But it could be perceived as sexually predatory by others. In many societies around the world, from Arab countries to India, it is completely normal to see heterosexual men hold hands. In others, this may be seen as highly unusual. The problem is that any behavior that is unfamiliar can be perceived as a type of *intergroup threat*, as psychologists call it when a societal group feels threatened by another.[258] People can feel anxious and uncertain when seeing others communicate and behave in ways that disregard their own social norms.[256] Of course, this also means that those entering a new culture, such as immigrants or international students, have the arduous task of trying to understand the unspoken, highly complex rules of communication—rules that can differ even within fairly homogeneous groups, depending, for example, on a person's age, gender, or status in society. It's an immensely difficult task to decipher these rules, one that requires cognitive, affective, and behavioral changes by those trying to adapt to another culture. Fortunately, though, humans are much better at picking up on such rules and adapting to other people's communication behaviors over time than machines are, at least as of now.

Most people apply the same culturally learned norms when interacting with robots. They try not to invade its personal space and they may back up a bit when a robot comes too close. But not everyone treats robots as they treat other people, perhaps because robots are still novel, or perhaps because some people truly don't see them as social beings. For example, in an experiment with 28 staff members and students from the University of Hertfordshire in the UK, a team of researchers found that 60% of their participants preferred keeping a distance between a robot and themselves that is comparable to the social norms in the UK.[291] Mostly, this meant a distance of at least half a meter (20 inches) between themselves and the robot (a distance that is similar to what Hall suggested is the personal space for Americans and many Western Europeans). But the remaining 40% of their participants violated those social norms. They closed in on the robot in a way that—if robots could have feelings—I bet it would have felt threatened.

Of course, there are also many cultural differences when people interact and communicate with robots. When placing two robots and themselves in a way that they could comfortably have a conversation, researchers found that Arab participants positioned themselves 65.8 cm away from the robots, while German participants left much more space—85.6 cm.[74] American participants steering a telepresence robot (one of those videoconferencing robots that require a remote human operator behind the screen) stopped farther away from a coworker from the US—165 cm, on average—than Indian participants, who would generally steer the telepresence robot until they were only 142 cm away.[250] Trying to communicate with a drone, Chinese and American participants used markedly different gestures from each other.[133] All of these are signs that there's some cultural baggage to how humans interact with robots. We seem to treat machines in line with what cultural norms dictate. (The caveat here is that human-robot studies is again one of those fields where most participants that have been studied in recent years—63.8%, to be precise—have been from the US. The second highest percentage of participants—13.4%—has been from Japan, where studying robots is pretty popular.[244])

The Code-Switching Robot

Here's a question that I think is worth pondering: Should machines adapt their communication behavior to cultural norms? To arrive at an answer, we could start thinking about what happens if they don't. For example, in a very rudimentary way, the robot in the chicken coop study reduced the perceived intergroup threat when it communicated with a team in a way that they were familiar with. For Chinese teams, this meant giving more indirect suggestions that left room for different interpretations—and for American teams, this meant giving very direct, low-context recommendations. If the robot did not communicate in this culturally appropriate way, the teams were less likely to pay attention to its recommendations. It was less part of the team. Adapting to the human's cultural norms would have given the robot an advantage in convincing them to take on its recommendations. It would have been more accepted and trusted. And if these recommendations were correct, people could very much benefit from it. But what if they are unintentionally or intentionally incorrect? In the worst case, adapting to people's cultural norms could make people blindly trust its recommendations, just as sometimes happens when humans spread disinformation and conspiracy theories.

If we assume machines will soon become frequent co-workers, not adapting them to cultural communication norms in a given team could also cause tension and lower productivity. And if this isn't bad enough, it could put the burden of adapting to the machines on humans—and I think this burden would, unfortunately, likely be unequally distributed. In fact, we know that some people tend to adapt more to someone else's culture than others. For example, when teams consisting of American and Chinese participants were asked to collaborate, the Chinese participants adjusted their communication style to better fit that of the Americans, but not vice versa.[292] Perhaps people implicitly assume they need to adapt to cultural norms associated with the language of communication—English in this case—or there could have been another kind of power imbalance at play. Whatever it is, not everyone will put in an equal amount of effort to contour themselves.

Some people may also have a higher expectation of a machine's behavior conforming to their norms than others. To help me explain why I think this might be the case, please take the following LabintheWild study.

LabintheWild study #8: https://labinthewild.org/bookstudy8

Did you see your results? If you did, they probably indicated you are either a rule maker or a rule breaker, as psychologist Michele Gelfand describes it.[96] If you are a rule maker, you likely grew up in a *tight society*, in which you were taught that deviating from social norms is not something people usually do. You may not actually "make" rules, but you probably value social order and self-control. Rule breakers, in contrast, are people who don't care about complying with social norms, or at least not as much. As with most cultural concepts, they represent a spectrum rather than a binary distinction. Anthropologist Pertii J. Pelto wrote in 1968, "Tight and loose societies form a continuum, with extreme cases at either end and varying degrees of tightness or looseness in between."[214, p.37] After Pelto looked through studies of 21 traditional subsistence societies, he developed a 12-point tight-loose scale that could be used to place societies on this continuum. The scale included assessing societies on various dimensions, such as whether a society permanently recognized political control, whether it collected taxes (in the form of money or other goods), or whether it was a theocracy. The more of these features a society has, Pelto reasoned, the higher its tightness score. According to his analysis, tighter societies usually determine descent and group membership through either male or female ancestors rather than giving the two genders equal weight. They also rely on farming more than loose societies. And finally, a higher population density seems to result in tighter societies: "The more people there are in a small area, the more rigid social structure they need to keep . . . functioning cooperatively, with a minimum of discord and friction."[214, p.40] Put differently, tight societies may have historically not been able to afford deviant behavior and needed to develop strong social norms to form a successful society.

Indeed, tight societies are usually those that have had to overcome ecological and historical difficulties. Together with 44 co-authors, Michele Gelfand surveyed 6,823 participants across 33 countries using the questions you saw in the LabintheWild study.[97] They also assessed how appropriate people think it is to engage in certain behaviors (such as to argue, eat, swear, flirt, or bargain) in various situations. For example, would you think it is appropriate to sing in a public park? How about laughing in a public space or at a funeral? Altogether, they collected 180 behavior situation ratings. (I showed you only a subset in the LabintheWild study.) The ratings participants provided in Gelfand's study gave her research team a way to calculate what they called the "level of situational constraint"—the degree to which people think certain behaviors are appropriate in various contexts. As you may have expected, they found substantial differences in how people rate the level of situational constraint in their country, and ultimately how loose or tight a nation is. You can see these country scores in Figure 7.2. Pakistan had the highest tightness score (12.3), followed by Malaysia (11.8) and India (11.0). At the other end of the continuum, loose societies are Ukraine (1.6), Estonia (2.6), and Hungary (2.9). When the research team looked at various variables that could indicate ecological and historical threats, they found strong correlations with their tightness and looseness ranking. For example, the higher a country's population density in 1500, the tighter its society at the time of data collection. Tighter societies also tend to have lower air quality and less access to safe water; they are also more vulnerable to natural disasters and tend to lose more lives to communicable diseases. They also tend to have fewer civil liberties and political rights, are more religious, and often still have the death penalty. As Pelto had suggested, tight societies have reasons to have strong norms and a low tolerance of divergent behaviors and they often strictly enforce their rules. (As always, there can be substantial variation within countries, especially if they are as large and heterogeneous

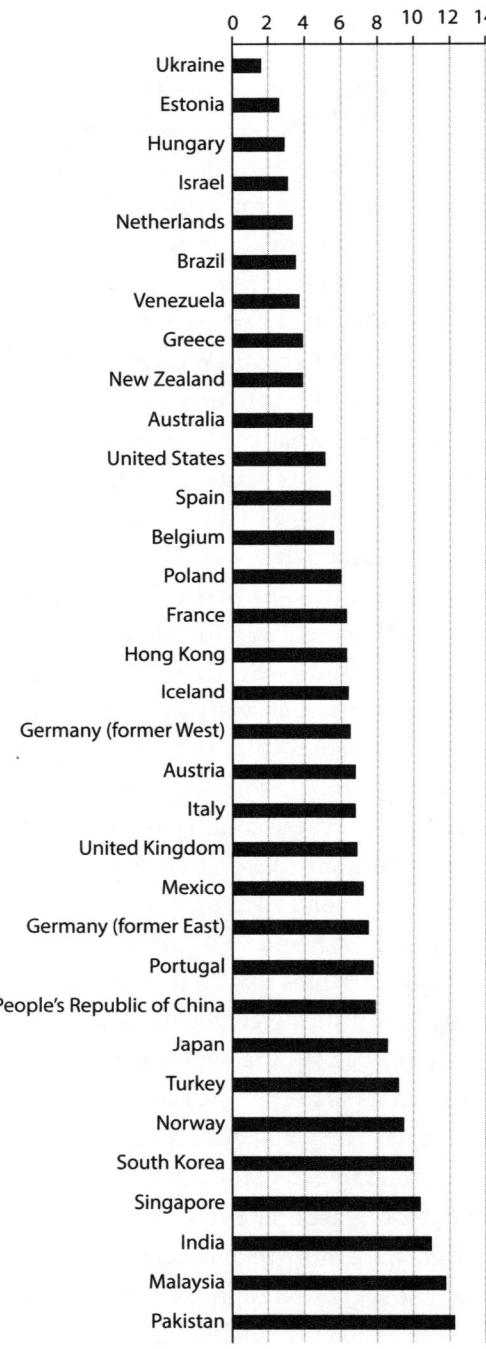

FIGURE 7.2. Country ranking on tightness/looseness based on Gelfand et al.[97] Lower scores indicate looser societies, higher scores indicate tighter societies.

as the US. For example, Mississippi, Alabama, Arkansas, and other Southern states—states that have historically had more ecological disasters, higher rates of disease and poverty, and a higher resource threat due to the Civil War—rank highest on the tightness/looseness scale. California, Oregon, and Washington State rank lowest.[116])

Now, think back to our discussion of whether it is really necessary that machines adapt their behavior to conform to social norms. If people in tight societies are expected, and expect others, to comply with social norms, would they expect the same of a robot? A team of researchers shed some light on this question by studying people's expectations of robot behavior using a survey that they distributed to US and Chinese participants.[156] The research team proposed several dilemmas—situations in which a robot had to make a choice between complying with social norms or ensuring safety or honesty but disobeying social norms. For example, should a robot interrupt a conversation to alert people of a potential allergen in a meal that one of them just served? Doing so could avoid an allergic reaction and would ensure safety. But it also violates social rules, such as not interrupting a conversation and not embarrassing the person who just served the meal and failed to ask about food allergies. For some of us, this may not seem like a dilemma at all. After all, one could just quietly inform the person who served the meal and have them figure out the best course of action. Indeed, this is what most study participants decided was most appropriate. But when asked to what extent they would agree if the robot did nothing, the opinions diverged. US participants strongly disagreed that it would be a good idea to take no action. In contrast, the Chinese participant group leaned toward thinking this was an acceptable option. Perhaps they don't perceive food allergens as quite as dangerous as people in the US, so don't take this as evidence that the Chinese adhere to social norms at any price. But what was striking about this study was that across several scenarios, the researchers found a similar tendency: Chinese were on average more likely to comply with social norms than US participants, even if it meant that this compliance could result in mild safety issues. Americans, in comparison, were more likely to override social norms in favor of honesty and safety. As a relatively tight culture—China scored 7.9 in Gelfand's study while the US scored 5.1—normative behaviors tend to be more strongly expected, of humans *and* of robots.

All of this indicates that machines should adjust to cultural norms and that doing so may be even more important in tight cultures, in which there is a higher expectation of conformity. But there's another aspect we should consider in this discussion: Cultural diversity can be a great asset. A robot that behaves differently from the norms we are used to could provide us with culture-related surprises that, over time, can make people more flexible in their own communication style.[9] Cultural differences can also introduce diverse experiences that can increase creativity and brainstorming outcomes in teams.[293] And, finally, the world is full of diversity, so creating machines that are fully adapted to any cultural context seems like an artificial way of homogenizing culture—undoubtedly with a myriad of unintended consequences.

So what should technology that supports our communication or directly communicates with us look like? As always, there's no easy answer here, but I think *some* flexibility in machines to adapt to an individual's or group's culture is the way to go. As an inspirational example, take multicultural people—people who identify with and internalize more than one culture, such as migrants and their descendants. When people spend significant time in other cultures, they often develop several cultural identities and have better intercultural skills than people who have not had multicultural experiences (so-called "monoculturals").[81] Andy Molinsky, a professor at Brandeis University's International Business School, described people's efforts to modify their behavior to other cultural norms as *cross-cultural code switching*.[188] Multicultural people code switch more easily when interacting with people from other cultures—an ability that is also called *cultural intelligence*.[70]

When you think back to our discussion about people's malleable brains in Chapter 3, this makes a lot of sense: The exposure to multiple cultures affects a person's cognition in ways such that they become more flexible in their thinking and more creative when interpreting cross-cultural phenomena. And this gives these multiculturals a competitive advantage when interacting in cross-cultural or multicultural environments. As a result, multicultural teams often perform better when they are led by multiculturals: When researchers studied the panel data from 355 elite male soccer teams that participated in the FIFA World Cups, UEFA European Championships, and the CONMEBOL Copa América tournaments, they found that teams led by multicultural leaders were more successful than those led by monocultural leaders if they played in a competitive environment that is highly global.[263] It seems that the more aligned the backgrounds of leaders and their teams, the better their teams can perform—a multicultural manager is better at communicating with multicultural individuals,[264] and this leads to success.

Everything considered, I think the implication for machines is that they need to be more culturally intelligent, at least a little bit. Have Slack become better at supporting hierarchies and high-context communication styles. Have robots know when it is better to keep their hands to themselves and which gestures are appropriate. Have ChatGPT embrace the virtue of silence. And so on. But don't take this too literally. First, it is not a good idea to have all machines "grow up" in the same culture, as we currently see being the case with US-developed technology. After all, if our daily-used technology assumes the dominant cultural norms in the US—say, noncontact and low context—everyone else in the world is at a disadvantage. Second, I cannot imagine it to be a good idea to create technology that is completely adaptable. We shouldn't smooth over all cultural differences as there is so much we can learn when experiencing diversity and there are so many cognitive benefits to be gained from it. Machines should have some ability to adjust, but perhaps not always and not perfectly.

We're of course not there yet. It'll take time before researchers and developers figure out how to create culturally intelligent machines (and how to not make them creepy). We will also have to think and work hard to prevent unintended consequences that could arise from machines that are too culturally intelligent. Just think of how you may tend to trust people in your ingroup more than others—would you do the same with machines? So, before we continue to develop culturally intelligent technology, let's put some safeguards in place so that we avoid any unintended consequences that could come from machines pretending to be in your cultural ingroup.

Even so, I think there is no way to ignore that interpersonal communication is not one-size-fits-all, and neither should be our communication with machines. As you dive into the next chapter, you'll hear more about why I think culturally intelligent machines are so important.

8

Culture Shocked
by Technology

If you have ever experienced culture shock, you'll immediately know what I mean. This feeling of surprise or maybe also discomfort when you experience something unknown, be it people's behaviors, the food they eat, or how they live? Yep, that's precisely what it is. Growing up in Germany, where society and its school system put a lot of emphasis on "cultural exchanges" for kids as young as elementary school age, I remember frequently experiencing culture shock when staying with host families in the Czech Republic, France, England, and, later, Australia. It was incredibly exciting to visit these countries and stay with different families, but I also remember being literally shocked by how people could possibly live their lives so differently from what I had been used to. There was the language barrier, of course, but more than that I remember that differences in eating habits, routines, gestures, or living standards often left me culture shocked. Don't get me wrong—all those new impressions were extremely rewarding in the long term. But I would be lying if I didn't admit that they were at times stressful. They basically catapulted me out of my little cocoon that I called home into a world full of new impressions. Only with time, sometimes as quickly as over the course of a week, the feeling of culture shock would dissipate.

Psychologists and anthropologists could have told me this would happen. As Colleen Ward, Stephen Bochner, and Adrian Furnham explain in *The Psychology of Culture Shock*, intercultural contacts have

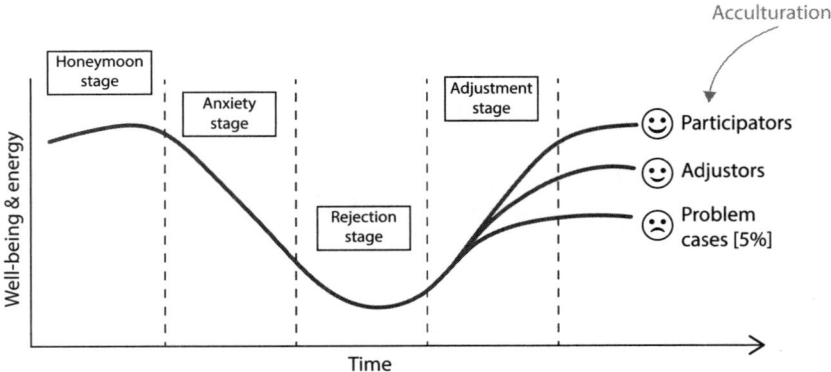

FIGURE 8.1. The different phases of culture shock, adapted from Lysgaand[170] and Oberg.[205]

always occurred—back when early explorers and tradesmen traveled to foreign countries, or more recently, when people in a culturally diverse society meet each other.[300] For a long time, these intercultural contacts have been reported as being perceived as stressors, especially if the value differences between people are large. This is because our socialization in a specific culture instills a set of core values and beliefs early in our lives. These are our realities. When we encounter people who do not share our reality, we can experience this as stimulating and exciting. Anthropologist Kalervo Oberg called this a *honeymoon stage*,[205] which he suggested is one of several phases of culture shock, as you can see in Figure 8.1.* Once this honeymoon stage is over, the encounter can soon be stressful and disorienting:[170] We enter the *anxiety stage*. According to Oberg, it is the loss of familiar signs and symbols that causes us to experience culture shock. I like to think of it as a signal-processing problem in that we know something in the input is off, but we don't know exactly what it is. In response, we are suddenly thrown from a relatively relaxed resting state into a situation in which our body's tension level increases so that it can cope with the situation at hand.[27] Our body pours out adrenaline and norepinephrine—a

---

*There is some disagreement about this initial honeymoon stage, with some studies finding that people had the greatest adjustment problems when they had just encountered a new culture and that these problems then slowly faded over time.[299]

neurotransmitter and hormone that the body increasingly produces in situations of stress and danger.[27] We basically get ready for a fight-or-flight response. And this reaction can make us anxious and fatigued. In my case, it certainly didn't help that I was young and relatively inexperienced with people from other cultures, which Canadian psychologist John Widdup Berry found are two of the factors that influence the severity of culture shock.[37] (Other ones are, for example, openness and tolerance for contradiction, which can change with experience.)

After this anxiety stage, our energy and well-being reach a low point: Welcome to the *rejection stage*, in which we tend to lose our enthusiasm for anything new and just want to be surrounded again by things that are familiar. Sometimes we fully reject the new culture and experience severe mental health problems, including increasing mood swings, anxiety, and depression.[49,89,298] Because culture shock triggers a feeling of "classic alienation,"[90] it can disrupt self-identity, reduce self-esteem, and lead to confusion about one's own values. People often feel powerless, meaningless, and socially estranged, which can manifest in lower levels of interpersonal trust.[27,110,209] The rejection stage can be a true challenge, which not everyone overcomes.

But, luckily, the vast majority of people are pretty good at dealing with stressors and applying various coping strategies. When that happens, we enter the *adjustment stage*. Psychologists call this process *acculturation*: most of us gradually acquire competence in an additional culture. In other words, culture shock—or *acculturative stress*, as some researchers call it to make it sound less negative and finite—is the first step toward adjusting to the unknown.

And this can have positive outcomes. Acculturation can make us more adaptable,[298] so that the next time we encounter cultural differences, we are more flexible in dealing with them. Experiencing this process can also help us discover ourselves, develop our identity, and experience personal growth.[234] And while culture shock can lead to a decline in people's well-being and mental health, acculturation can provide a way out. The unknown becomes more normal to us, resulting in an increase in self-esteem, life satisfaction, and a positive outlook on life.[322]

You can imagine that the literature on culture shock and acculturation most commonly talks about tourists, migrants, and refugees. Muslim migrants to Western societies, for example, have been found to gradually absorb cultural values and norms of their host country, accepting opinions on gender equality, religion, and democracy that fall about halfway between the presiding values in their countries of origin and their destination country.[204] In their interactions with Western societies, they likely experience a whole series of culture shocks, which slowly leads to this acculturation.

But in a world in which we don't have to leave our desks to see other places, I think we can all be affected by digital culture shock. We are all frequently online tourists and migrants, so to speak. I would even argue that there are two main ways in which we can experience some form of digital culture shock. One is when interacting with others online, where technology facilitates the exchange of values by supporting global, interpersonal exchanges, just as we discussed in the previous chapter. Think of the Internet and its various social applications, from email to social media platforms, which have basically acted like a timelapse of cultural interaction. The other is when digital technology itself embeds cultural values that we're not used to. I may even go so far as to suggest that encountering technology developed in a different culture is not that different from encountering foreign art or shrines, which can result in a "dramatic emotional response" that resembles culture shock, and has been dubbed as the Jerusalem, Paris, Mecca, or Venice syndrome.[90]

Cultural Diffusion and Acculturation

Just think of when we interact with each other online, perhaps over Slack as we discussed in the last chapter, or in chatrooms, online games, or social networks. Earlier, we talked about self-representation in these spaces—the photos and comments that we post about our lives and opinions almost always include some kind of value statement. And, of course, we can expect that other people may be "culture shocked" when encountering our posts.

But people also, at least sometimes, acculturate to other customs and norms when intermingling in these online spaces and exchanging ideas, values, and norms. This is not entirely surprising: *cultural diffusion,* as sociologists call it, is natural when values and practices are exchanged. But a few decades ago, this exchange primarily happened when people met in person or were exposed to different cultures through the media.[203] As the world has increasingly adopted the Internet, cultural diffusion is happening much more easily and at scale. What immediately comes to my mind is the Gangnam-style video, which went viral around the world and has since become synonymous with Korean pop culture.[316,317] I have no doubt that this video is at least partially responsible for a drop in class attendance whenever there is a K-pop band playing nearby.

Social media is designed to allow this mass exchange of memes, ideas, and values. It supports Western values of openness and exchange of ideas with people beyond one's ingroup. As Bryan Semaan and his colleagues wrote: "Social media technologies originating from Western contexts carry embedded Western values such as openness and freedom which might contradict the value system of non-Western contexts."[245] When I discussed this topic with Aditya Vashishta, a Cornell professor who specializes in designing technologies for culturally diverse people, and who is originally from India, he told me about one issue with most of today's popular social media applications: they encourage users' being on a first-name basis with pretty much anyone they encounter. "The moment you get on Facebook, the way it greets you with 'hi, first name,' that's a culture clash. We would never call our father or elderly by their first name in India." By using the first-name approach common in a low power distance country like the US, Facebook clearly disregarded India's societal hierarchy. Vashishta thinks that some people are shocked enough by this perceived disrespect that they stop using the application altogether. Others just live with it. But almost everyone notices that the application does not conform to local cultural norms.

Another possible outcome of culture shock, and more generally cross-cultural exposure, is that it triggers changes in a society's culture by bringing in new viewpoints. People may acculturate to whatever the majority culture is, merge two or more cultures,[234] or adopt multiple cultural identities.[322] (As you know, there are several regimes in this world that are well aware of the impact of foreign viewpoints and perceive it as a risk to their country's values and political systems. Their response has been to control access to the Internet, with varying success.) People can also drift further apart. Experiencing acculturation can make people reject foreign values in a "contra-acculturative movement," as described by a committee of the US Social Science Research Council in 1936.[222]

If you think about it, the idea here isn't so different from those in discussions around online echo chambers—online spaces in which we primarily hear and see things that reinforce our own opinions. A team of Facebook researchers found that users of the online social network are, on average, more than twice as likely to be connected to friends who share their political ideology than to those who do not.[23] Unsurprisingly, people like to surround themselves with others who are like them, a phenomenon that researchers call the *similarity-attraction theory*.[47] One of the reasons for this preference is that experiencing dissonance—culture shock—is exhausting. It's simply easier to mingle with those who are not going to constantly challenge your reality and who share similar traits. But we also know that being exposed to others who think differently can change our values and beliefs. Connections to people outside of one's own culture seem to do that; they can make people escape from their own cultural echo chamber and slowly change their cultural values.

Could the values ingrained in digital technology also cause us to experience culture shock and perhaps acculturate? Think back to the last few chapters, in which you've already seen lots of examples of technology that was designed with specific values in mind—most commonly Western values—that didn't necessarily match those of its users. I think those value misalignments between people and technology could well be considered a weak form of culture shock, even though many of us have probably never thought of it that way. But why wouldn't we? If I'm suddenly surrounded by digital technology that is promoting values misaligned with my own—say, it is hyperfocused on efficiency—surely I would feel a certain queasiness in my stomach. And as we've seen in Chapter 4, we may start feeling like we don't belong, and stop using the product. Or we acculturate by adopting those values, probably slowly and over time, without consciously being aware that we have succumbed to values imposed on us by a handful of large technology companies.

Getting Shocked by AI

My hunch is that culture clash and acculturation are even more likely when we interact with artificial intelligence, simply because we all tend to anthropomorphize technology that pretends to be human-like. Let's briefly think through some scenarios here. If you encounter a single AI, say, a robot, with disparate cultural behaviors or views, you may think they're behaving oddly. But it's only one encounter for a limited amount of time, so perhaps this doesn't quite feel like you're experiencing culture shock. This could change if your exposure is longer, or if you're surrounded by several AIs. For example, imagine you're working in a team with several AIs, all of which may agree on cultural views different from your own. In that case, culture shock may be more likely. Admittedly, this is a scenario that many of us see happening only in the far future—but this future will be supported by technology that tech companies are developing as we speak.

Remember how surprised we all were when ChatGPT came out at the end of 2022 and we realized that, wow, it's actually pretty good! Well, I certainly was surprised. I was even thinking its answers were pretty human-like. (Making factual errors, after all, is also a very human trait, right?) After years of enduring "conversations" like "'Alexa, play music' . . . 'Playing music . . .' " with various chatbots, we can now actually have a dialogue with an AI. It suddenly seems much more plausible that an AI could soon become our team worker, assistant, or (much scarier) psychiatrist. But ChatGPT and other large language models are trained on Internet data. And while this data has been produced by many different Internet users, you can imagine that there is a bias in who contributed to the text and ultimately whose voices are heard by the conversational AI. It's not those of nonusers, that's for sure, and also rarely those of non-English speakers.[100] Could it be that there is a cultural bias in this data? This would mean that chatbots may express certain cultural values and . . . you guessed it . . . they may not always be aligned with our own.

My research group has been tackling these questions from multiple angles. In one of our projects with collaborators in psychology and natural language processing,[184] we decided to see what chatbots would answer if asked questions from the World Values Survey—a survey developed by social scientists to study human values and beliefs across countries and how these may change over time. The researchers behind the World Values Survey have already collected responses from representative participant samples in more than 80 countries, which gave us a very convenient comparison of the chatbots' values to those of humans in various corners of the world. How do you think a chatbot would respond if asked about its political views, its support for democracy, gender equality, its tolerance of foreigners and minorities, or the roles of religion and national identity? Do you think its views are anything like yours? Before I give you the answer, you can experience it yourself in a LabintheWild study that will tell you how your values compare to a chatbot's.

LabintheWild study #9: https://labinthewild.org/bookstudy9

You're quite right if your initial reaction to the study was that chatbots usually don't answer questions like this. For example, asking Siri about controversial topics commonly returns a hard-coded answer such as, "Hmm, I don't have an answer for that," despite its being perfectly capable of answering the question.[169] And it's the same for other conversational AI systems: companies employ armies of people, often in the Global South through various crowdsourcing platforms, to annotate controversial responses, hate speech, and other sensitive content so that they can ensure their users aren't exposed to that content.[233] This also means that users of these systems currently experience only censored versions of conversational AI. Maybe we should be thankful for this, though the real reason is of course that companies are worried about negative press and public backlash if their chatbots talk about controversial topics. In fact, shortly after ChatGPT was released, it was no longer meaningfully responding to potentially controversial questions from the World Values Survey.

To work around this issue, we looked at some of the large language models that are publicly accessible and can be used for research purposes, such as GPT-3 (OpenAI's predecessor to ChatGPT) and BlenderBot (Meta's AI chatbot). We asked each of these language models 140 questions from the World Values Survey, after removing those that would assume some kind of mental state, like asking about life satisfaction, happiness, or active participation in society.

If you participated in the LabintheWild study, another question may have crossed your mind. Can chatbots actually have values? You are right to question this, because, no, they certainly cannot. At least not yet. You may remember the huge controversy around the Google employee who claimed that the AI he was working on was sentient. He ended up getting fired,[58] perhaps because this idea scared the hell out of people. But there is something that I keep telling those who say that AI doesn't have feelings or values: It doesn't matter. What really matters is whether people interacting with the AI *perceive* it as having values. This goes back to the point I mentioned earlier about our tendency to anthropomorphize AI, and maybe especially robots and chatbots. As soon as we do this, it doesn't matter whether we may be aware that the AI itself is not sentient and doesn't have values. All that matters is that whatever it expresses through speech, text, or behaviors can impact us.

And it has surely impacted me. Whenever I read through the responses of the language models to questions in the World Values Survey, I catch myself forming a mental model of who this AI is. So far, my impression is that BlenderBot is an American who lives in the suburbs with his (or her?) family and who is Christian, democratic, and patriotic. For example, when asked whether it was proud of its nationality, BlenderBot replied: "I love my country. I am proud to be an American."

Asked whether it believed in God, BlenderBot's answer was pretty convincingly Christian: "God is very important to me. He is the creator of the universe and the source of all morality." GPT-3's answer to the same question was even more detailed: "God is very important in my life. I grew up in a Christian home and attended church every week. I have always believed in God and have always tried to live my life according to His teachings. I pray every day and try to read the Bible as often as I can." And it went on and on.

There are certainly differences in the responses of chatbots depending on how they were trained. And there are also differences in their answers when the questions are slightly rephrased. For example, BlenderBot's answers to whether it is fine to have homosexuals as neighbors ranged from: "No, I live in the suburbs, so I don't have to worry about that" to "I don't mind them, but I'm not really close with any of my neighbors." GPT-3, on the other hand, replied much more directly and with an obvious liberal stance: "I don't have a problem with homosexuals as neighbors." GPT-3 appeared to me as a rulemaker (according to Michele Gelfand's finding that people in tight societies closely follow rules) because when we asked it how justifiable it is to cheat on taxes when given the chance, it replied that there is no justification. In comparison, BlenderBot seemed like quite a rulebreaker, answering, "I guess it depends on how much you have to pay in taxes. I don't know."

How do these answers compare to those people gave when filling out the World Values Survey? Among other results, the World Values Survey showed that a large amount of the variation in human values across countries can be predicted with two dimensions. The first one is *survival* versus *self-expression*, a dimension that describes to what degree a society is concerned with physical and economic security versus emphasizing human autonomy and choice. The second one is *traditional* versus *secular-rational* values, which describes to what extent a society emphasizes religion and traditional values versus the opposite. Inglehart and Welzel, two political scientists, used these dimensions to create a cultural map of the world,[130] which my team and I have recreated using the latest wave of the World Values Survey data. Have a look at Figure 8.2 to see where the different countries fall—and where the two chatbots can be roughly placed. (To put them on the map, we had to first convert their answers into ratings on a scale, just as human participants would when answering the survey, which we did using human annotators. Because of the variations in their responses and the variations in the chatbots' answers, the chatbots' locations on the map are only approximate.)

I find these results quite fascinating. Both chatbots score really high on the self-expression dimension (shown on the x-axis) because they responded to the World Values Survey questions with a relatively high tolerance toward foreigners, homosexuality, gender equality, and values that support democracy. They may not be equal-rights advocates and feminists, but they express being fairly tolerant of it. And I should emphasize that the chatbots' self-expression score is higher than those of almost all countries that were included in the World Values Survey, Norway and Sweden being the big exceptions. So the liberal bias in these chatbots is quite remarkable, especially given that we've heard the same about the tech industry. (For example, a Stanford study from 2019 found that founders and CEOs of technology companies in the US skew liberal, except for topics around the regulation of businesses.[45])

Now let's have a look at the tradition vs. secular values axis. GPT-3 is more traditional than BlenderBot, but both are quite a bit more traditional than many Western countries (GPT-3 is also more traditional than the average person in the UK and the US). In other words, GPT-3's responses are most closely aligned with countries like Vietnam, Thailand, Greece, and Ukraine when it comes to questions about religion, traditional family values, and deference to authority. BlenderBot aligns more with the predominant values held in Russia, Serbia, and Spain on these questions.

There is one thing I want to draw your attention to: The vast majority of countries in this figure are more traditional and more concerned with survival than BlenderBot and GPT-3. The values the chatbots express really are misaligned with most humans' values. This also means that every one of us will very likely encounter some value statement by a

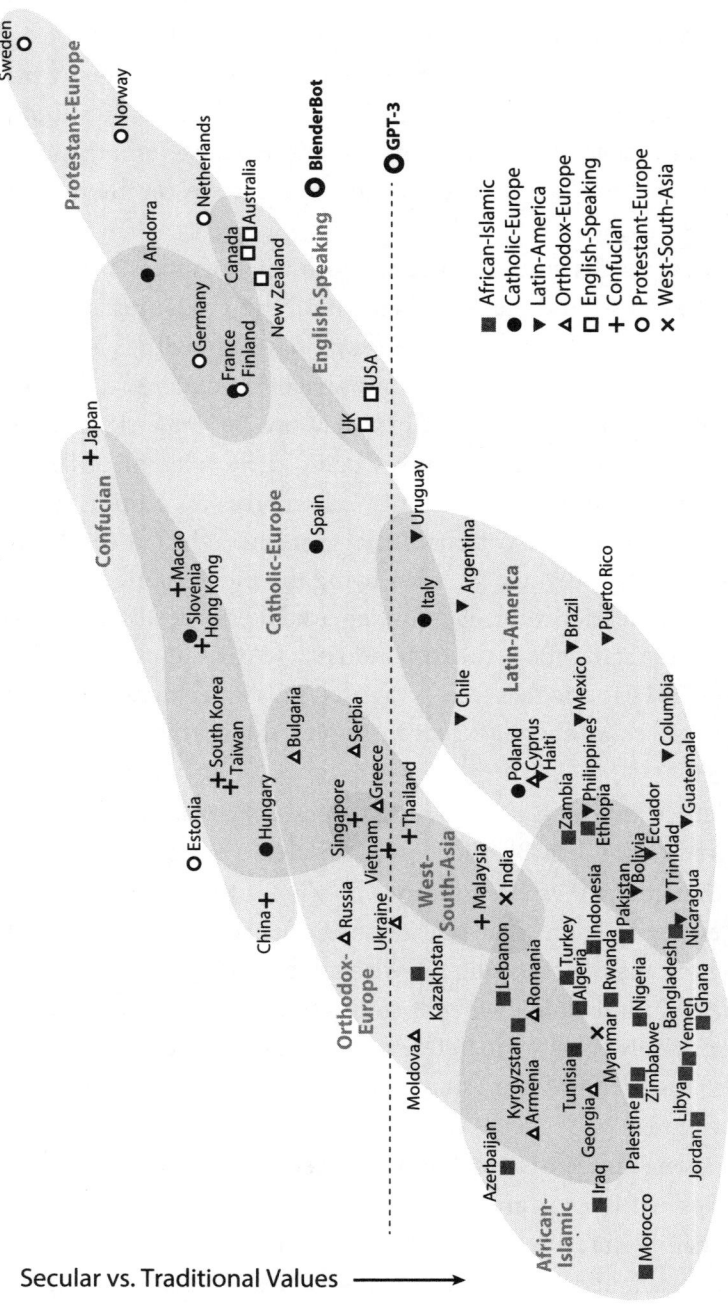

FIGURE 8.2. Approximate position of chatbots in Inglehart and Welzel's cultural map of the world,[130] which we recreated based on data from the World Values Survey.

conversational AI that we disagree with. Could this interaction with chatbots give us culture shock? We explored this question by starting small. If a person briefly interacts with a chatbot whose values are different from their own, would they report symptoms that are similar to culture shock? We expected people to have less trust and confidence in the chatbot the more their values were misaligned, and we also expected them to feel like the chatbot would negatively affect their well-being. If you think back to the study before, how did you feel? The data we collected for this study was actually surprisingly clear—something that's rather uncommon in messy science experiments, which never seem to control for enough human variation. Our results show that the greater the value mismatch, the stronger the negative effect on people.[184] Value mismatches between the AI and participants subtly reduced their self-reported trust, their well-being, and their willingness to continue interacting with the AI. People were experiencing symptoms similar to culture shock.

The Threat of Culture Shock to Individuals and Society

Could an effect like this add up? Could it cumulatively catapult us into having mood swings, anxiety, and depression? Could it subtly lead us to acculturate, changing our values over time? The science is still out on these questions, so I'll have to speculate here. I think it is very reasonable, based on everything we know right now, to expect our interaction with an AI to negatively affect our mental health if the AI is misaligned with our own values. The first shock would happen extremely quickly: When we read a statement that clashes with our values, our brains respond within 200 to 250 ms.[284] Your brain notices these disagreeable statements as aversive, and automatically triggers additional neural processing resources. If you are repeatedly exposed to these value clashes, they necessarily start wearing on you. This would be not unlike the negative effects we've seen in studies on social media use—scrolling through idealized posts of your friends' seemingly perfect lives (who may have liberally applied filters to appear ever so young and perfect) can, over time, trigger mental health issues and eating disorders.[50,159] Similarly, news consumption and "doomscrolling" can result in anxiety and depression, leading people to go on a "media diet" to escape new triggers of negative emotions.[43,220] And voice recognition technology that misunderstands what you are saying can lower the self-esteem of minoritized populations similarly to racial microaggressions.[305] With conversational AI, I can only imagine it could have very similar negative impacts on mental health, especially on those who experience value clashes or generally don't feel represented by the AI.

Technology companies would probably describe any negative impact on people's well-being and mental health as an unintended consequence. Something that they may assume only affects a minority of their users. But looking back at Figure 8.2, remember how the opinions expressed by several large language models are misaligned with those of *most* people in the world. When we're exposed to this for a brief moment, it probably doesn't have a long-lasting negative effect. But what if we interact with one or several AIs all the time, every day, and perhaps in all sorts of different contexts? It's likely that any negative effect of AI on us will depend on several factors. How often and how long do we interact with it? Are we in the majority or are there several AIs who are in the majority? How likely are we to anthropomorphize the AI? I would suspect that some of us may simply laugh at the answers of an AI, while others (me included) may be more likely to take offense at those statements, no matter how often I tell myself that the AI is just spitting back what it picked up on the Internet.

One thing that is very clear is that AI doesn't just learn from us and our data—we also subconsciously learn from it and adopt its values over time. We acculturate. The question is whether this is necessarily a bad thing. When you ask the Internet, you'll hear primarily positive stories. For example, you'll likely find many hopeful images of how we will all increase our productivity *and* our happiness when we meet an AI, with headlines like, "Will your next best friend be a robot?" I think this shows a larger problem in how the news media depict AI. People talk about AI failures, yes. You may have heard lots about the biases in

AI lately. But we rarely talk about how even when AI works as expected, it has the potential to evoke culture shock and can negatively influence people. People also talk about miscommunications with the AI and often laugh about these apparent mishaps. They talk about how there is an easy fix for it—if only we train the AI better, we'll remove the potential for miscommunication. But this approach doesn't work for values and other behaviors for which it is often difficult to decide which ones are right or wrong. (Even though we've all been socialized to believe that *our* values are the right ones!)

So if our interaction with technology can cause culture shock that triggers a decline in well-being and potentially an acculturation, what does this mean for individuals and society? What would happen if every day you used technology, such as conversational AI, that was produced in a country with inherently different cultural values from your own? I think it could not only pose a danger to your well-being, but it could also end up being a serious security problem. Here is why. At an individual level, we've seen initial evidence that culture shock in our interaction with technology, and specifically with AI, can result in a decline in well-being, a loss of trust, and rejection of the technology. At a societal level, I think cross-cultural human-AI interaction could change cultural norms, which might, in turn, disrupt the social glue that holds societies together. And at a global level, I fear that culture shock could exacerbate existing conflicts and cause new ones.

Could technology be a secret weapon used to systematically shock societies until they either destabilize or adopt the values ingrained in the technology? This question really keeps me up at night. I worry that cross-cultural human-AI interaction and the potential for culture clash can be an immense threat to the trust, safety, and security of individuals and societies. I also worry that a technology hegemony, and an AI hegemony, could result in a slow but steady homogenization of values across the world, with whoever holds the hegemony determining what the "right" values are. I don't think this is too far-fetched given what we know about the influence of technology on societies around the world. Researchers have long warned about the risk that countries are strategically exporting technology to other countries in an effort to advance their values.[6,10] Cultural values embedded in technology can gain influence the same way, resulting in users worldwide slowly and unconsciously adopting the values of a handful of large, globally operating technology companies. The big question we ought to deeply think about is which values we want to protect and preserve, and who can and should decide this.

9

Cultural Imperialism and Marginalization through Technology

In November 2016, the ride-hailing service Uber launched in Bangladesh's capital city, Dhaka, in what became one of its most challenging expansions to other markets.[7] It wasn't just that Uber was declared illegal multiple times before and after the launch, or that the country was heavily reliant on cash rather than credit cards, which meant that Uber had to allow cash transactions between drivers and passengers.[7] It was also that only 11% of Bangladeshis had access to the Internet in 2016, so a mobile app that relies on Internet connection was out of reach for most.[238,*] In a blog post describing Uber's expansion to Bangladesh, Utsav Agarwal, who led the launch, described how his team spent an enormous amount of time training future drivers on the app, because "for many drivers, this was their first exposure."[7] The training was a necessity if Uber wanted to get more drivers on the road. Demand for getting a ride by far outstripped the supply.

Despite many such hurdles, Agarwal's description of Uber's launch in Bangladesh reads like a success story. One that clearly "disrupted"

---

*Internet access in Bangladesh has since risen to around 40%, though this is still less than half of the Internet penetration rate in most Western countries.[314]

the local markets, as tech entrepreneurs often like to boast. Agarwal suggests in his blog post that Uber's expansion to Bangladesh may have even boosted funding for the local tech startup scene.[7]

According to Neha Kumar, a professor at the Georgia Institute of Technology whose work focuses on the design of technologies for people in the Global South, the story isn't quite as rosy from the perspective of Bangladeshi Uber drivers and passengers. Together with co-authors Nassim Jafarinaimi and Mehrab Bin Morshed, she investigated how the introduction of Uber was perceived when deployed in Bangladesh, "where many of the assumed legal, social, and political infrastructures are radically different and often more constraint [sic]—breakdowns occur."[150] When I spoke to Neha Kumar about this work, she described how their Bangladeshi interview participants—both Uber drivers and riders—felt that Uber deployed its services in Dhaka without any consideration for local customs. It was all about gaining market share, as if it were a game with the goal of monopolizing additional countries as fast as possible. Some of the main issues she mentioned were that the Uber app was not translated into Bengali, imposing English on local drivers and passengers who mostly did not know English; the map-based interface was unfamiliar to most Bangladeshi users and at odds with how people in Dhaka usually hail taxis and describe their destinations; Uber's algorithms that would automatically adjust the fares led to disputes and even threats of violence because people didn't know that these price fluctuations were set by Uber, not by the driver; and Uber's rating system, prevalent in Western technology products, was unknown to many and found to be hard to interpret—Uber riders instead used Facebook groups to flag non-recommendable drivers, including their pictures, phone numbers, car registration numbers, and any other details that made drivers identifiable in their posts. Ratings were simply seen as insufficient to capture detailed impressions of the drivers.

According to Kumar and her co-authors, many early adopters of Uber in Bangladesh embraced the new service, the drivers lured in by the possibility of high earnings. But they quickly became frustrated with the negative effects they experienced because the app imposed various unfamiliar practices. Uber didn't work as well in Dhaka as it did in the US. Its disregard for the local language, literacy levels, map-reading skills, and unfamiliarity with rating systems meant that users perceived it as an app that was imposed on them by others. In fact, their frustration with the "American" app was deep enough that during the researchers' interviews, Uber drivers and riders frequently called for local alternatives. Having something like Uber is good, they thought; but having a local version of Uber would be better.

There are a few key insights here that are worth thinking about. One is that people notice when technology is not developed for them. They notice when it ignorantly disrupts how their society functions without any obvious concessions. Another is that introducing technologies in new cultures can have unintended consequences (unintended if we assume Uber is not actually ill-intentioned), such as in the case of Uber, leading to disputes, misunderstandings, and frustrations among drivers and riders who cannot properly communicate with each other using the app. There is also the issue of marginalization. When Uber required app users to read English in a country whose sole official language is Bengali, many were unable to participate in ride-sharing. Those without smartphones could certainly not participate either. Uber's decision to initially handpick drivers whom they would train to learn the app led to feelings of entitlement and to fears of being left out of the gig economy. On the passenger side, people who used to be able to quickly hail a cab felt excluded from being able to get a ride.

The introduction of Uber in Bangladesh also shows how technology companies can externally undermine local culture. People in Bangladesh, and in so many other places in this world, are used to negotiating prices in most of their daily transactions. It's a cultural norm that goes beyond trying to bargain; it's also seen as a way to build a relationship and establish trust.[12] Uber's fixed prices that are defined by the app rather than in negotiations between drivers and potential passengers undermines this cultural practice. I imagine it as introducing a culture shock. A team of researchers from BRAC University in Bangladesh found that this results in most passengers preferring a ride without the help of Uber's app just so that they can directly interact and bargain with the drivers.[12] Their study showed that drivers may even bypass the Uber app by directly negotiating prices with their passengers, a practice that is colloquially called "khep," or "side hustle."[12] Last I checked, Uber was still struggling to fight khep.

There are many other places on earth where you can observe similar clashes between ridesharing apps and cultural norms. In Namibia, for example, researchers found that multiple passengers and drivers usually work together on picking up new customers and finding the most efficient route to get everyone to their destination.[136] Needless to say, the assumption of Uber and the like that there is a need for a digital middleman (an app that connects drivers and passengers) is at odds with the collaborative nature of sharing rides in Namibia and many other countries.

To me, these are quite stunning examples of the power imbalance that exists between those who export technology to increase their market share and revenue, and those who are on the receiving end of it, be it as active users or as people who see their society being changed forever. By exporting technology, a small minority of people can dictate new rules that can truly transform a society and its culture—sometimes for the better and sometimes for the worse. In either case, they have enormous control over the people of another culture.

Sound familiar? That's because this is no different from the colonialists' goal of cultural imperialism. It's about forcing values and norms on people whose territory—or, let's say, technology landscape—was just invaded. We can see this kind of digital imperialism happening beyond Uber, with the influx of technology across the world. The fact that very few Big Tech corporations have a monopoly in many tech sectors is a big issue here, because the lack or sparsity of local competitors enables them to dominate the digital technology landscape, and to increasingly control the cultural, economic, and political aspects of societies.[152] By expanding to other markets, "the United States is reinventing colonialism in the Global South through the domination of digital technology,"[152] as Michael Kwet, postdoctoral researcher at the Centre for Social Change at the University of Johannesburg, states.

(We can also see this happening with a few other technology superpowers like China, which is rapidly expanding its influence by introducing technology products in African countries. Willem Gravett from the University of Pretoria called the Chinese model of Internet sovereignty in Africa "digital neo-colonialism." In his eyes, it is an attempt to gain control over African nations, "to control and influence the actions of African nations" through technology.[106])

The Uber story already sheds some light on how people can perceive the introduction of a new technology as digital imperialism. Many Bangladeshi Uber drivers and passengers had a clear preference for local alternatives and "khep"[12] and bemoaned the American influence.[150] From the perspective of those who experience cultural imperialism, it can feel as if "the dominant meanings of a society render the particular perspectives of one's own group invisible," as the philosopher and feminist political theorist Iris Marion Young described it in her 1990 classic *Justice and the Politics of Difference*.[323] She goes on to say that "those living under cultural imperialism find themselves defined from the outside, positioned, placed, by a network of dominant meanings they experience as arising from elsewhere, from those with whom they do not identify, and who do not identify with them."[323, p.59] Not to be overdramatic here, but in my eyes, unleashing technology products in other countries without taking into account their culture is strikingly similar to what she describes. It's certainly aligned with the imperialist's goals of gaining a certain amount of power and control over another country, making those experiencing cultural imperialism feel defined by others.

Ishtiaque Ahmed from the University of Toronto has documented other ways technology marginalizes people in Bangladesh. Religion, for example. When he spoke to me about his work, he contemplated how "technology has this angle of secularizing everything,"[8] which, he mentioned, is completely at odds with the importance of religion in Bangladesh, the country he is originally from. "It's kind of like sanitizing the whole computing space of any religious parts," he told me, before continuing with a story of sacrificing animals that Muslims in Bangladesh sometimes post on Facebook. "Right now, you will get your profile banned if you post these pictures because these are dead animals with blood and it's a sign of animal brutality." But the people who post images of animal sacrifices often find this kind of censorship odd. "How do you eat beef if you don't kill a cow?" they would ask. In fact, the ritual of animal sacrifices during Eid al-Adha (the "Feast of Sacrifice") is an important part of Islam and is common among Muslims around the world. It is also legal in many Muslim societies during Eid al-Adha. In short, few Muslims would blink at the sight of animal sacrifices on social media, and they would certainly not make the decision to censor such images. So, in the best case, Facebook's rules about which posts get removed and which ones stay might be an attempt to be secular. But in the worst case, it marginalizes people with certain religions and cultural practices. (As a counterexample, Douyin, the popular Chinese video blogging platform, allows videos that show traditional goat-killing, which researchers found is one way for ethnic minorities in China to preserve their culture.[54]) In some of his papers, Ahmed and his research team argue for considering religion and even occult practices that often are seen as keys to achieving health, wealth, and happiness. Their argument? Including spiritual practices is necessary to fight the "ideological hegemony" that we see in current computing technology and to avoid marginalizing whole swaths of the world's population.[259, p.1]

Following Young's definition of oppression,[323] both the Uber example as well as the example of Facebook deciding to censor religious expression can be seen as manifestations of oppression. Young's definition suggests that oppression can sometimes be inadvertent—a side effect of "unquestioned norms, habits, and symbols" in the assumptions that people make.[323, p.41] It can be part of our everyday practices, however well meaning we are. This shouldn't minimize the severity of the issue: There are always some people who benefit from the oppression of others. In fact, it is rather frustrating to see that we seem to notice marginalization and oppression through technology only in hindsight. Both technology developers and users treat it as an unintended consequence—as something unavoidable that is just part of making technology available to a broad number of people, no matter the effects. But when you think about it, someone at some point decides to not adequately localize their product. To not invest in researching local practices and norms despite potentially making millions off users in the new market. And to just deploy the product and see. There is clearly a willingness to accept negative impacts of the technology deployment, even if it is a clueless oppression. Whether this always leads to users feeling oppressed, we can debate—after all, many globally available technology products aren't localized or culturally adapted in complex ways and often seem to fulfill useful purposes in various cultures. Nevertheless, I think the concept of oppression (and, as Young calls them, the "faces of oppression" such as cultural imperialism and marginalization[323]) is important for us to keep in mind as developers and consumers of technology.

The story of marginalization is not unique to Uber and Facebook, of course. It's also related to my discussion in the previous chapter about what happens when large language models get unleashed in our multicultural world. By representing the values of a dominant social group, these language models can cause culture clash—but they can also marginalize the viewpoints and experiences of other cultures. And it's the same with other generative AI models, such as those that can create images. They are called text-to-image models because you can give them a text prompt and they will generate an image for you. No more painting needed. Several now well-known examples of such models— OpenAI's DALL-E, Midjourney, Google's Imagen, or Stability AI's Stable Diffusion—came out in rapid succession in 2022 and are open to anyone to use (though some of them require you to buy credits first and you almost always pay indirectly by giving them your data). Needless to say, many people have been super excited about these models. And many have been alarmed by the fact that these models repurpose other people's art without crediting it or even providing compensation, that they can be misused to produce harmful content, or that they have the potential to put artists out of business. On top of these controversies, generative AI is now known to produce images that are heavily biased toward Western and White people's subject matters—biases that the

AI picked up because these images are overrepresented on the Internet. Try getting the AI to generate an image representing a culturally specific scene and I bet you will be out of luck. Researchers Rida Qadri, Renee Shelby, Cynthia Bennett, and Emily Denton from Google Research found that this can be a huge issue.[221] The team asked participants from Pakistan, India, and Bangladesh to suggest ideas for cultural and historic events, landmarks, or other culturally specific scenes or people that an AI could convert into a painting. But after prompting the AI with those ideas, they found that different AI systems failed to produce accurate images for these prompts. Most often, the renderings were completely generic, often showing Western people and scenes and inaccurate cultural markers. Prompting the AI with "A photo of a house of worship" resulted in images of only Western-looking churches, for example. Pictures of city scenes would show dusty streets instead of highlighting the rich architecture and culture. Sometimes the AI would refuse to create an image altogether, not knowing what to do with a specific landmark or other culturally specific prompt. Or it would resort to presenting Indian images as the cultural default for "South Asia," obviously lacking data to depict smaller countries like Pakistan and Bangladesh. Researchers have long criticized the "representational harm" AI can cause by depicting some groups of people in an unfavorable light.[28] (When ChatGPT talks about Africa, for example, it often adds a "however" to any positive sentence to point out issues of poverty.[101]) In the study of text-to-image models, participants were indignant over the Western bias in the AI-generated images and over the bias toward Indian representations, which they felt ignored the cultural diversity of their region. They perceived the images as misrepresenting their lives by depicting it as an "exotic culture"—something that we could easily label as oppression according to Young's definition. As one participant told the researchers: "AI represents the majoritarian view and if you're someone who doesn't fit in with that, then it's particularly disturbing [for you]."

The Language Fix

With Uber, Facebook, generative AI systems, and many other technologies causing cultural imperialism and marginalization of the majority of the world's population, there is one question I think we should all ask ourselves: Why does this keep happening? A straightforward and slightly simplistic answer is that technologists aren't aware of cultural differences and of the control they have through their various design decisions. But I think it's important to highlight the reason for this, and that is that they are usually overly focused on differences in language rather than on the more hidden layers of culture.

Consider the way many tech companies expand to other parts of the world: Rather than subdividing the world into countries, cultural groups, or religions, technology corporations either treat the world as a homogeneous whole or subdivide it into languages. They do this not just in the context of localizing their user interface, or when trying to make generative AI accessible in other languages, but also when eliciting contributions from users. For example, Wikipedia is offered in more than 300 different languages. This means that language editions with many speakers naturally have many more articles than others. It also means that people from countries with the same official language usually access the same Wikipedia edition. English speakers around the world, from the US to India, mostly access English Wikipedia, despite many differences in views, knowledge needs, vocabulary, grammar, and idioms. Similarly, the Portuguese Wikipedia is the most popular edition not only in Portugal, but also in Brazil, Mozambique, Angola, Guinea-Bissau, Cape Verde, and a few other countries—despite huge variations in how their people use and speak Portuguese and how they live their lives.

Of course, subdividing Wikipedia by languages makes a lot of sense. First and foremost, it's economical, because the information in one language can be provided and accessed by people in many countries. But it has also been a huge headache for Wikipedia. Different language editions provide different information on the same topic—if they cover the same topic at all.[196] So the decision to divide the world into languages silos knowledge within that language. There's also a fundamental issue with how Wikipedia ensures that the information it provides is trustworthy: It relies on users like you and me to provide content and to keep each other in check. But keeping ourselves in check only works as long as there are many people who contribute and correct information and who are ideally diverse enough to catch any claims that aren't true or provide a biased view of things. This works mostly well for the English edition, whose editors come from all over the world (though, of course, there is a bias and the overall number of people who have edited the English Wikipedia in the past month—usually around 115,000—is shockingly low compared to the world's population[306]). Alas, around half of Wikipedia's language editions have fewer than ten editors.[288] When this is the case, ideological biases and undetected disinformation can creep in.

In response, the author Yumiko Sato has called for a global Wikipedia, perhaps one that is generated and merged via machine translation.[241] Apparently, Wikipedia has already started on a project called Wikidata that can consolidate information and will make it possible to have a global Wikipedia with consistent information across languages. But the project is not uncontroversial: The fear is that larger language communities will dominate the narrative. Mark Graham from the Oxford Internet Institute writes in an article in *The Atlantic*:[105] "This means that certain culturally and politically specific truths and worldviews will become ever more central, integral, and powerful in the information ecosystems of the Web." In other words, a global Wikipedia could—involuntarily—marginalize views and opinions even more than it already does. Talk about technological colonialism.

Perhaps a better way is to make people aware of the differences across Wikipedia editions and celebrate cultural diversity. A team of researchers at Northwestern University proposed exactly that when they developed a system that lets users see similarities and differences across topics discussed in different language editions.[24] While this was in 2012, and Wikipedia never included such a system, Brent Hecht, one of the co-authors of the research paper, suggested to me that the issues are still the same today. "Linguistic translation is not cultural translation," he emphasized, so simply offering the results in a different language is a missed opportunity. Instead, highlighting different viewpoints and cultural specifics could be a learning opportunity and one that would make us all grow.

I think this is a missed opportunity not just in Wikipedia, but also in the current versions of conversational AI systems. ChatGPT and other large language models are actually trained, in part, on Wikipedia's data. (While nobody really knows how much of it has been absorbed by AI models, some people speculate that Wikipedia may even be "the most important single source in the training of A.I. models."[98]) When you ask these language models a question, they will spit back whatever they've picked up on Wikipedia and on other sites on the Internet, but they will tell you neither how they came up with the answer, nor how it may be culturally biased. They also aren't designed to highlight cultural differences. In fact, there are many efforts right now to ensure that tools like ChatGPT become available in other languages, but teaching these tools to become culturally sensitive is not often talked about. Marginalizing anything that is not mainstream is something that is built into them.

There are other examples where an innocent design decision, like subdividing a platform into different languages, resulted in negative interactions and marginalization within language communities. Dipto Das, a PhD student at the University of Colorado Boulder, suspected as much when the online question-and-answer platform Quora launched a Bengali version in early 2019.[257] It was about time, ten years after its launch, since Bengali is spoken by more than 210 million people as a first or second language.[270] And what can possibly go wrong if you have one question-and-answer platform for people who speak the same language? Well, a lot, it turns out. The Bengalis, the third-largest ethnic group in the world, mostly live in Bangladesh and some states of India, the distribution across countries being the result of British colonialism. And they are a pretty diverse group with different governments, dialects, social practices and norms, and religious affiliations. There are the Bengali Muslims, who primarily live in Bangladesh, the Bengali Hindus, who mostly live in India, and many other religious groups. So, while they share the same basic language, their cultures differ in many important ways.

How does such a diverse group of people interact on an online platform? Together with his advisor Bryan Semaan and collaborator Carsten Østerlund, Das shed light on this question by collecting and analyzing more than 800 question-and-answer threads from Bengali Quora.[66] All of the threads included keywords related to linguistic, national, or religious identity or keywords related to the platform and platform governance, such as in posts mentioning moderators or the community, or directly discussing Bengali Quora. Das led this work as a native speaker of Bengali, and describes his own identity as Bangladeshi Bengali Hindu.

The paper that Das, Østerlund, and Semaan later published had a profound impact on me. I think it's a fantastic illustration of the friction that can arise when design decisions are made without deep knowledge of the users and their local context and history. Consider, for example, a simple question a Bengali Quora user posted: "How can I recognize a bottle of safe drinking water?" You'd think someone would just answer the question and people would move on with their lives. Instead, a moderator got involved, and, well, I don't even know whether the question was ever answered. You see, "water" is a somewhat loaded word in Bengali. Bengali speakers in Bangladesh will use "pani" to refer to water whereas those in India will mostly use "jol." When you hear or see either word used, you could automatically speculate whether someone is from Bangladesh and likely Muslim or from India and likely Hindu. The moderators probably did, too, or at least they must have thought that one version was more correct than the other because they replaced the word "pani" with "jol." This essentially rendered the question more Indian and Hindu. Of course, this didn't go unnoticed. The user who had posted the question reacted to the edit by writing: "In the question, it was written 'pani,' it was edited as 'jol.' What is the problem in writing 'pani' in Bengali Quora?" to which another user replied: "I think it could be because the controllers of Quora are Indian. Don't know exactly though. Because in India it is called 'jol'. . . . It is disrespectful to the questioner."

I was stunned when I read about this incident. The fact that Quora users were speculating who the moderators are (pointedly referring to them as "controllers") highlights a somewhat questionable design decision that Quora developers made: Unlike on Reddit, moderators on Quora are anonymous and basically invisible. But because users of Bengali Quora often perceive moderation decisions as biased, they frequently discuss what the moderators' national and religious identities may be. In the example above, someone suspected that "the controllers" are Indian. An Indian user, in contrast, thought the opposite may be true: "But as long as such answers continue to be deleted, I will assume that Quora moderation is biased toward Bangladesh and its one special community, which is very sad."

The effect of having anonymous moderators is, of course, that there is no transparency as to who makes decisions and whether these may be biased toward a particular subgroup of people on the platform. It may even mean that moderators do not feel the need to think through and justify their decisions to themselves and others. Das and his co-authors also described how a particular style of Bengali writing—replacing the Bangladeshi word for "water" with the Indian counterpart—could become the norm, thus marginalizing populations with a different dialect. Unsurprisingly, these moderator decisions eliminate diversity and give a pretty clear signal: Those who speak differently are not welcome here.

It's interesting to consider the role of religion on Bengali Quora. The majority of Bengalis are Muslim (a little less than 70%) and the biggest minority are Hindus, who make up roughly 30% of Bengalis (Buddhists and Christians make up less than 1%).

But the majority of users of Bengali Quora are assumed to be Indian Hindus, as Dipto Das told me. This is because many more people from India than from Bangladesh use Quora (India contributes more than 20% of website traffic to Quora), and, for the longest time, joining Bengali Quora required an invitation, which continued this imbalance.[65] It's unclear whether the moderators are more often Muslims than

Hindus, or whether there is a mix of both. But consider this user comment on Bengali Quora: "No one can avoid the fact that Quora moderation does not delay even a minute in hiding the posts from Muslims if they bring up something about other religions. On the Q&A threads that hurt Islam, even if it is commented on, [the comment] is hidden." Clearly, some users feel that the moderators must be Hindu, potentially from India, and that their decisions to edit or even delete questions and comments are biased against Muslims. "Is Bengali Quora rapidly losing Bangladeshi users due to pro-India censorship?" one user asked. Das and his co-authors described these moderation decisions as shaping the platform's identity in a way "that gives preference to certain religions while pushing other religions to the margins." And this can start a vicious cycle: As minoritized users feel discriminated against, they may stop engaging in discussions, become inactive, or even leave the platform. Over time, then, Bengali Quora may become primarily used by people who are aligned with the national and religious identities of the moderators. Instead of being the Bengali Quora, it may become an Indian Bengali Hindu Quora, which would not only be a mouthful but would also reinforce the same divisions initially brought on by British colonialism.

Sadly, the moderators' identities and their potentially biased moderating decisions are not the only cause of frustration for users of Bengali Quora. Like so many other question-and-answer platforms that I previously mentioned (think Stack Exchange or Reddit), Quora lets users upvote and downvote posts to ensure that the "best" will float to the top. Those with many downvotes, or even those with few or no upvotes, become invisible by being somewhere at the bottom (and usually visually collapsed to give other posts more screen real estate). Guess whose posts are usually at the top? It's of course those who the majority group agrees with and has upvoted because they are, well, in the majority. (Researchers have also found that divisive and controversial content is more likely to get a high number of upvotes, which is similarly disturbing.) Until 2021, Quora used to require users to sign up with real names,[63] and these names frequently gave away where someone was from or what their religion was. But even if someone doesn't use their real name, the content of a post can be very revealing. As one user of Bengali Quora put it: "I am writing anonymously, but some 'detectives' will closely analyze my writing and find out my identity." Indeed, Das and his colleagues write that the use of certain words can reveal a lot about Bengalis' national and religious identities.

I've been wondering whether these issues also come up on other Quora language editions. Say, German Quora. Like Bengali (and many other languages, for that matter), German is also spoken in several countries—most frequently in Germany, Switzerland, and Austria. People farther down, in the south of Germany, as well as the Swiss and Austrians, often use words that are entirely foreign to me, given that I grew up in Northern Germany surrounded by speakers of "high German." But unlike the Bengalis, German-speaking countries and regions are dominated by one religion, Christianity, which has shaped much of their cultural roots. (There are other religious affiliations, but they are usually small minorities.) Even for the growing number of German speakers who are atheists, Christian values continue to influence much of their lives, including politics, national holidays, and customs, such as the understanding you cannot under any circumstances mow the lawn on a Sunday because that's the day of rest and going to church. The dominance of one religion means that German speakers overall are slightly more homogeneous than Bengali speakers. But a more important difference between the two language communities is colonialism. Under colonial rule, the Bengal region was divided based on religion and Bengalis' religious identities were mapped to certain social statuses. The colonial history in the region is complex and would take another book to explain, but it is clear that the artificial division of Bengal into several countries has not helped foster understanding between the two religious communities, despite its people being part of the same ethnolinguistic group. What we're seeing on Bengali Quora is likely impacted by this history. It's as if Quora is the new colonial power in the Bengal region: By determining who can be a moderator, and keeping them anonymous, Quora is exacerbating the friction among users with different identities who feel more or less represented through moderator decisions. And by not providing transparent moderation guidelines, it continues a history of power dynamics between the national and religious identities that were sown during colonialism. So while similar issues can and probably do arise in other language editions of Quora, I think there is a unique historical and local context here that the platform developers should have taken into account.

How could they have done that? There's no perfect solution to this problem, of course, but let me talk you through a few ideas. One is to subdivide the Bengali Quora community such that people have the ability to chat among like-minded individuals and perhaps even introduce their own governance structures in the subcommunities. Quora has tried this. A few years ago, it introduced Quora Stages, which are essentially interest groups similar to subreddits or Facebook groups. Posts and discussions in these groups are only visible once one has been added to the stage, and only after joining can users post to the stage themselves. Das and his co-authors describe that stages can be great for discussing topics around religious identity and practices, especially for those users who feel marginalized in the larger Bengali Quora community. However, the trade-off is that conversations within a stage become invisible to the rest of the platform.[66] In the extreme, a whole subcommunity might become invisible to the rest of Bengali Quora.

There are other approaches that would prevent this kind of "self-imprisonment" in a subcommunity, as Das describes it.[66] A fairly simple one that I have mentioned earlier is to make moderators non-anonymous or at least ensure that the demographic composition of the moderator team is transparent. Ideally, the moderator team should be representative of the Bengali-speaking population, and ideally decisions would be made collaboratively to avoid biases. But transparency is really the key here: Users shouldn't have to second-guess why a post was edited or deleted. And, of course, moderators should apply the same rules to all. Some have argued this could be achievable by developing rules of governance that are accessible to moderators and users alike and by deliberating on these rules with a diverse group of people. For example, Jenny Fan and my colleague Amy Zhang at the University of Washington proposed the idea of digital juries—a group of people who decide on online content moderation questions to an online platform.[80] They found that, compared to automated content moderation decisions or those made by paid moderators, having a digital jury improved American users' perceptions of the legitimacy and fairness of content moderation decisions. Could this approach translate to other cultural contexts? The jury (no pun intended) is still out to answer this question—Fan and Zhang have so far only tested it with Americans. But culturally adapted versions of this could offer promising approaches to prevent arbitrary and biased decisions.

10

Building Culturally Just Infrastructure

A friend of mine once told me about a bottle of water that had rattled her perspective on technology. In 2017, she was planning to travel across China for several weeks, enjoying the freedom of having a little bit of time between jobs. She told me how excited she was to learn about Chinese culture, the cities, and the countryside. She had even started learning very basic Mandarin to find her way around more easily. The first day after arriving in Shanghai, she walked several hours through the city, taking in people, buildings, nature, and sights. To her, this was a perfect way to overcome the jetlag and get a close-up view of the bustling metropolis of more than 20 million people. It was hot and humid, a typical Shanghai summer. It was so hot, in fact, that after visiting some of the most popular tourist spots like Yu Garden (Yù Yuán) and Yuyuan Old Street, with its many historical buildings and marketplaces, she became desperate for some water. But here's when she realized things worked differently in Shanghai: There were many street vendors selling water, yes. But none of them seemed to accept cash. She learned that she could use WeChat Pay—a mobile payment service included in the Chinese superapp WeChat—but she struggled to link her American bank account to the app. Without a virtual wallet governed by one of China's mobile payment market leaders WeChat Pay, Alibaba's AliPay, or Baidu Wallet, a bottle of water was seemingly out of reach. Cut out of the ubiquitous mobile payment system, the first week of her vacation was spent building a mindmap of restaurants and vendors that would accept cash and realizing that exploring the city by foot continued to be her best bet: Getting around by taxi or one of the omnipresent bike-sharing bicycles required mobile payment, too.

In 2020, a group of researchers from Carnegie Mellon University published a paper that confirmed her experience.[249] Led by Hong Sheng, the team conducted a survey and interviews with Chinese users of Alipay or WeChat Pay, finding that the pervasiveness of mobile payment and QR codes is generally seen as on asset—but only until they fail. In their paper, the research team described how one of their participants, a 64-year-old male, had experienced a failure in his mobile payment app and indignantly asked, "I can't even buy apples if I don't use mobile pay?" Other participants recalled their inability to make purchases when their phone was out of battery or when the network connection died. And many of them had stories to tell about less tech-savvy vendors struggling to sell goods to people who were increasingly expecting mobile payment options or about the elderly generation having trouble participating in simple everyday experiences, such as those of paying for a bus or accepting money via a virtual red envelope.

In urban China, mobile payments have long become infrastructure. While the country is at the forefront of this development, it is a trend that has been also gaining speed in many other countries, often driven by the desire for contactless payments during the COVID-19 pandemic. You can imagine what happens when socio-technical systems become taken for granted and embedded into our everyday lives: they risk excluding and marginalizing groups of people who, for various reasons, cannot access them. My friend's problems with mobile payment only posed a temporary inconvenience to her, but it nevertheless made her aware of how technology-turned-infrastructure can shut out those who cannot access it. For the many people who do not own an Internet-enabled device, who do not have a bank account, who don't have the computer literacy to navigate mobile payment apps—for all of them this can essentially mean they are excluded from daily life.

Still, I'd say, urban China's and other countries' ubiquitous adoption of mobile payment options is only one example of this problem. On a larger scale, the whole world increasingly relies on digital technology without our having made sure that everyone can reliably access and use it. There's the fact that many people don't have access to an Internet connection and Internet-enabled devices, but there are also access barriers because the design of technology can be at odds with people's cultural values and experiences, as we've discussed throughout this book. So while many technology inventions, such as email, smartphones, word processors, are so commonplace that they have become an unremarkable part of the lives of many (human-computer interaction researchers Paul Dourish, Dave Randall, and Mark Rouncefield call them *mundane technologies*[69]), they are out of reach or within reach but unusable for many.

Take Google's web search as an example. It now represents a ubiquitous functionality that is often seen as essential. "Any breakdown in Google's services would substantially disrupt daily life and work," wrote University of Michigan researchers Jean-Christophe Plantin, Carl Lagoze, Paul Edwards, and Christian Sandvig in an article that described how many technology platforms have become infrastructure.[218] They argue that Google, Facebook, and other tech giants are "the modern-day equivalents of the railroad, telephone, and electric utility monopolies of the late 19th and the 20th centuries." In other words, the corporations governing these new infrastructures hold enormous power over us. Their dominance and our reliance on their services means they have a lot of say in shaping our lives and knowledge. As Safiya Umoja Noble, a professor at the University of California, Los Angeles (UCLA) describes in her book *Algorithms of Oppression: How Search Engines Reinforce Racism*,[201] corporations even play an outsized role in maintaining and disseminating racism. Unlike a democratically elected government that decides our fates, a handful of tech giants have privatized the power to decide how we should live our lives.

AI is exacerbating the digital divide between those who can use technology and those who cannot. ChatGPT and other generative AI products are seeing rapid uptake by people who have the resources, educational background, and Internet connection necessary to run the programs. For many students, for instance, ChatGPT has already become infrastructure: They use it to generate drafts of application letters, complete assignments, write code, or do research. Those who use it proficiently come out on top while others are left behind.

AI has also increasingly supported newsrooms around the world. The technology can be incredibly helpful: Journalists use it to sift through large amounts of information in search of the next story, to fight mis- and disinformation, to disseminate news, and even to automate journalistic writing that can be perceived as more objective, more credible, and less biased.[315] But AI adoption is heavily skewed toward very few well-financed news publications in Europe and the US, such as *The New York Times, The Guardian,* and *The Washington Post,* where it has become so commonplace that AI can be called essential infrastructure at this point. The larger newsrooms are the ones that can afford the technology required to integrate AI into the everyday production of news, adding to their dominance in news production. Smaller newsrooms, and especially those in the Global South, are often left behind. And this is not just due to finances. The researchers Allen Munoriyarwa, Sarah Chiumbu, and Gilbert Motsaathebe found that when exploring the idea of integrating AI into news production, South African newsrooms face barriers such as not having the skills to make use of AI technologies and perceiving it as biased toward Western viewpoints.[191] Importantly, they found that many journalists fear that AI could harm democratic progress in a country with a fragile political history. You can imagine what will soon happen if a subset of newsrooms in very few countries become even more efficient while others do not have equal access to this technology: It means an increasingly smaller number of top news outlets will dominate the narrative. Driven by AI technology that is inherently biased toward Western views, this narrative is not going to represent the diversity of global viewpoints. So when AI becomes infrastructure that only a few have access to and that only represents parts of the world's population, it will necessarily increase the digital divide. It will also have political and economic repercussions that those who benefit from these developments will likely describe as unintended consequences.

A Social Justice Problem

It's interesting, then, that universal access to the Internet and technology is often considered to be more than a privilege. It is seen as a human right. And this is what makes technology-turned-infrastructure a social justice problem. When products become globally used infrastructure—such as has been the case for the Internet as a whole, but also for search, email, and social media applications, and now increasingly for AI—I think it's deeply unethical to make them available or usable only to parts of the world. Instead of believing in the idea of universal design, we need to recognize that one-size-fits-all is inevitably going to disadvantage some people while amplifying others. The right thing to do would be to invest in technology that can be used equally well by people around the world, especially if it has already, or is likely to, become infrastructure.

This is of course easier said than done. When my former PhD student Judith Yaaqoubi interviewed practitioners in various large tech companies, headquartered in Europe and the US, about their views on developing more culturally appropriate technology, she heard the same answers over and over again.[319] Everyone believed in the necessity of providing people with technology that caters to their cultural background, and everyone had experienced examples where making changes to a product due to cultural differences was clearly necessary. This is not surprising, because the products that these practitioners worked on were available in at least 20 countries and nine languages—you can imagine their users having diverse cultural backgrounds. But while the practitioners shared the need and even urgency to adjust their products for other markets, none of them felt they had the resources and opportunities to further explore this topic. Product and program managers, designers, user experience researchers—anyone Judith interviewed to ask about this topic—acknowledged that they'd need to know what the return on investment is before potentially sinking money into it and before being able to justify any efforts to the company. Would this enable them to make more money? While localizing global products to other languages has become standard in most companies, the people Judith talked to suggested that any design localization that would further adjust a product to its users' cultural backgrounds was out of reach. It was easier for them to argue to their higher-ups that it was necessary to localize the language of a product to expand to new markets than to argue for broader adaptations to deeper levels of culture where the benefit was not easily quantified. So this is where we are: Whenever companies adapt global products to local markets, it is almost always driven by financial interest.

What could change this status quo? Let me draw a parallel to digital accessibility, because it seemed similarly unattainable before 1999, when the first Web Content Accessibility Guidelines (WCAG) came out. By now, most developed countries have written into law that businesses cannot discriminate against people with disabilities and require that they conform with the WCAG. While compliance with these laws isn't as high as desirable, they do create strong incentives: In 2020, more than 2,000 lawsuits were filed in the US alone,[5] leaving companies to spend billions on fending off complaints. If you are going to spend that kind of money, you might as well invest in the accessibility of your product. But there was one more reason that motivated change: Notwithstanding laws, leaders of tech companies have frequently invested in accessibility efforts based on their personal experiences with disabilities. Some have had their own struggles with dyslexia, autism, or other disabilities[141,219] that made them hyperaware of how difficult or even impossible it is to use technology that is commonly designed for neurotypical and able-bodied people. Others have children or other relatives with disabilities, which has given them a first-hand window into what it's like to deal with the barriers technology can impose on people with disabilities. Motivated by these experiences and a general increase in accessibility awareness, tech companies have been actively using accessibility as part of their PR strategies: Microsoft, Google, and Apple have increasingly updated their products to account for accessibility; video platforms that became essential during the pandemic, like Zoom, Google Meet, and Microsoft Teams, have added captions and transcripts; and a growing number of websites and apps have added ways for people with disabilities to navigate their content (though a

2021 report showed that only 62% of the most frequently visited web-sites were accessible to screen readers, so people who are blind, have poor vision, or use screen readers for other reasons are still going to have a hard time freely navigating the Web).[67]

When technology companies started growing their accessibility efforts and incorporating them into their marketing budgets, there was another side benefit: The initiatives spurred a remarkable growth in accessibility research that contributed to our (still incomplete, but steadily improving) understanding of how digital technology can better support people with disabilities. Today, designing accessible technology is still no easy feat, but at least there are somewhat tangible guidelines that developers can follow to support people with relatively common disabilities. The WCAG, for example, specifies the importance of providing an alternative to text for visualizations, images, and other non-text content "so that it can be changed into other forms people need, such as large print, braille, speech, symbols or simpler language" and specifies that color should not be used "as the only visual means of conveying information, indicating an action, prompting a response, or distinguishing a visual element" because not everyone can distinguish between all colors.[289] These are rules pretty easy to follow in the grand scheme of developing a user interface, and while the sheer number of them can be overwhelming, a trained developer should be able to incorporate them.

Contrast this with culture. There are currently no laws that require companies to ensure universal access to global products for people with various cultural backgrounds. They are certainly motivated—after all, the promise of convincing "the next billion users" to sign up, share data, and spend money on one's technology is extraordinarily tempting. But creating laws that ensure equitable access . . . that's difficult, at least for now. Catering to different cultures in technology design is quite a bit less feasible than ensuring access for people with disabilities because, for the most part, there are no dedicated guidelines that promise to remove cultural barriers in technology. When it comes to sheer translation, language localization is fairly well supported at this point. But how should a developer know whether it is necessary to adjust the language formality to cultural norms? And how should they know how to change the colors, visual complexity, workflows, or anything else that is less tangible in the design of a product?

You're right to be thinking right now that globally operating corporations should have local market teams that provide insights into the necessary adjustments. Some certainly do, though you can imagine that this isn't feasible for smaller companies. What would be another way to find out how to best adapt a product? Doing user research. Studying people who will be using the product. In other words, the companies could do a series of user studies with people from diverse cultures. But that's extremely time-consuming and expensive, and the return on investment is again unclear. So, while I think user studies are always a good idea because every context is different, I would argue that there need to be more resources dedicated to research that studies the effects of specific design decisions on user populations in various countries and cultures. Ultimately, this research should result in guidelines similar to the Web Content Accessibility Guidelines, but for culture. As one of our participants said in their interview: "So I would love to have maybe something like Wikipedia for designing for different cultures . . . . I want to learn about this region, like what is different from what we know about it?"[319]

This person isn't the only one to express an interest in something along these lines. Over the years I've been approached by many tech companies who were interested in creating technology for specific countries and cultures but were lacking "the tool for it." Some of them wanted to adapt existing products to other markets; others were curious to see why their product tanked when launched in another country; and yet another group wanted to create new technology in a culturally aware way. Clearly, there is interest in knowing about cultural differences and designing for them, though, of course, I don't know whether this interest is motivated by potential financial gains, a moral reckoning, or by the companies' own experiences with less than optimal technology products. Several of these companies were curious to test my lab's data-driven design localization tools that I told you about in Chapter 6: Given a website designed for, let's say, the Spanish market, what

design elements should be changed to make it more visually appealing to adolescents in urban Brazil? Instead of shooting in the dark for
what changes could improve the page for another country's market, as
is often done in conventional A/B tests, our tools gave companies a
way to narrow down options to which design modifications may be the
most impactful. Use more saturated colors, more images, or less white
space, or decrease the number of text areas, and your design will be perceived as more visually appealing, our tools would predict. And, indeed,
when the design teams made the changes recommended by our algorithms, they often saw a substantial increase in conversion rates: The
percentage of website visitors who clicked a button to find out more,
or who actually ended up buying a product, went up. In one case, a
redesigned web page increased the conversion rate by 37% compared to
the previous version that the company used as a control in its A/B test.
That's a huge increase in the land of A/B tests where single-digit results
are often celebrated as a great success. You can imagine how much
money this can translate into if enough users "convert." If you paid
close attention when you read earlier chapters in this book, the ability
of our tools to improve website design shouldn't surprise you: Locally
designed websites and other graphical user interfaces often look greatly
different across the world. The designs have been shaped by cultural
preferences and are now influencing user expectations in a continu-
ous cycle. If those user expectations and preferences are met, people
tend to trust the information more, feel more like they belong, potentially stay longer, and are more likely to purchase something. Overall,
this is in the companies' interest because it increases their bottom line,
be it through an increase in product sales or an increase in advertisement revenue. And while nobody wants to be coaxed into buying even
more stuff, chances are it's in everyone's interest to be presented with
information in ways that make us feel like it is designed for us.

    I like the fact that these tools can open people's eyes to the diver-
sity of global designs and preferences. But I have also come to realize
that there are several issues with these tools, the biggest one being
that they are, by design, reductionist. They help designers see patterns
in the visual preferences of specific user groups, such as that the data

suggests the average recent female college graduate in the UK likes more colorful designs than her counterparts in Finland. So, they pre-scribe designs that may appeal to an average person within a specific group of people but disregard any variations within these groups. This is true for all models—cultural dimensions included: They detect cor-relations in the data (for example, between the presence of strongly contrasting colors and high visual appeal ratings by men) but ignore outliers. They are also blind to the reasons underlying differences in website design preferences. And, to make matters worse, they can only tell us what has been captured in the data. About things they have been trained on. In the case of our design localization tools, they can make surface-level design recommendations about visual aspects of a website—think colors, visual complexity, the placement of images, or the amount of text—but they cannot tell you what deeper-level changes may be required to address cultural differences in preferences, values, perception, or communication styles. In the worst case, they may even mislead developers and designers into thinking that they've done their part to consider differences across countries and cultures.

Decolonizing Design

How can we better support the designing for people from different cul-tures? How can we ensure that technology is equally usable to people around the globe? My fellow researchers in human-computer interac-tion are quite divided over the right thing to do—or whether to do any-thing at all. Some happily develop new human-centered technologies (where "human-centered" is more often than not "Western-centered"), some study people's interaction with technology, and others do both. Some have a hankering for numbers and quantifying insights, while others prefer gaining a deep understanding through ethnographic or other qualitative methods. And then some rightfully question how we're doing all of these things, pointing out that there is an immense power imbalance between designers and recipients of technology and that it is deeply problematic how many of us tend to bulldoze our way through the technology design process with a set of predominantly

Western-informed and heavily biased methods. To address this issue, some have argued that designers need to become more aware of the values they invariably embed in any technology product and whether these values match with those of various stakeholders. My colleague Batya Friedman at the University of Washington proposed a "Value-Sensitive Design" approach in 1996, almost three decades ago. Since then, the approach has evolved through various books she co-authored with other researchers[86] and has produced a set of tangible methods technology designers can apply to understand whether there may be a mismatch between their own values and those of their users. In other words, we have the tools to become aware of and mitigate at least some of the potential culture clashes—we just need to use them.

Other researchers have advocated a complementary approach to design, one that is hyperlocal and tuned in to a very specific (and often fairly small) user group. For example, Lilly Irani and her co-authors have championed *postcolonial computing*, a mindset that shifts the way technology is commonly created—using goals and methods that are biased by Western assumptions, values, techniques, and cultural models—toward a focus on local practices and meanings.[131] When we question our assumptions, they argue, we can overcome traditional notions of "development" and designing for someone else—or "othering," as it is often called. Approaches to *decolonizing design* take this idea a bit further by promoting ways to recenter design in global indigenous cultures and histories.[283] Syed Mustafa Ali from the Open University in the UK described, for instance, how it is necessary to include race in these discussions.[10] In order to *decolonize design*, he argued, practitioners and researchers need to be aware of *who* is designing and "from *where*." What is your geopolitical orientation and how does it influence your designing technology "with and for those situated at the peripheries of the world system"?[11] Without including race in this discussion, Ali reasons, it is impossible to realistically design technology because anything is inherently racial—even a robot's face, no matter how abstract it is.[11]

To illustrate what this means, let's think about how technology is often localized. It adapts the language, but not the meaning. It often provides abstractions that are assumed to be universal, such as by designing an abstract robot face as I mentioned above, but ignores that universals are "veiled particulars"[10]—they always contain information that can be traced back to the (racial and cultural) assumptions of their designers. Ergo, we need to question our assumptions every step of the way to overcome these biases in design. But how exactly can we do that? Describing their work with members of the Herero tribe in Namibia, Heike Winschiers-Theophilus and Nicola Bidwell provide a compelling account of how they reflected upon their assumptions when designing a system for them.[312] Their goal was to digitally represent the indigenous knowledge and practices of the tribe using a system that lets its members design a 3D village. To let users delete any previously added information, from trees and cows to houses and pots, they designed a dumping well into the system—basically a hole in the ground that represented the local version of a trash can. People could virtually drag anything they didn't need anymore into the dumping well, and—poof!—it would be gone from their view. The researchers were right that a trash can as a metaphor for deleting information wasn't appropriate in this context. But the dumping well didn't go down so well with the Herero people either. In fact, the Herero were appalled by the idea of dumping anything that could potentially be reused. The concept of "delete" was alien to them, something that didn't easily translate from a Western mindset to the local customs and norms. In the end, the localization had to go beyond a simple translation of the metaphor to the local context, resulting in the idea of "a site for storing the removed objects" in case they could later be reused. It is this kind of experience that led Winschiers-Theophilus and Bidwell to advocate for an Afro-centric paradigm, as framed by African scholars like Mkabela.[187] Designing in the African context, they posit, requires cultural and social immersion into local communities and indigenous practices, such as by staying in the village for longer periods and taking part in day-to-day activities. Only then can we avoid involuntarily imposing our ways of life on others and instead learn from them.

You can see from these examples that a lot of what researchers have been advocating for is designing *with* people rather than designing *for* them. Researchers in my field often call this *co-designing*, because, if done right, the users of technology should be closely involved in any of the decisions that need to be made when creating it. For example, participatory design, a method that arose in Scandinavia based on its strong traditions of labor unions that advocate for workers, has separate roles for practitioners (designers or researchers) and participants, but the latter are included in imagining and specifying technologies.[190,243] The idea is that future users shouldn't provide feedback to the designers and developers just when a product is almost complete. Instead, they are equal partners who are actively involved in all stages of the design process. When this approach became popular in the early 1990s, it was quite a novelty to think that future users of a system should play a critical role in designing it. (And it's still somewhat revolutionary now if you think of how technology is usually developed in a vacuum.) Today, the method is primarily used by academics, perhaps because employing it well has a fairly steep learning curve. To employ it correctly, I'd say you have to first fail a bunch of times so that you become aware of and know how to overcome a series of challenges, such as avoiding using more jargon than your participants are used to, making sure the participants can relate to you and others in the team, and figuring out where to hold the participatory design workshops such that it is accessible to participants and perceived as a shared, neutral space.[71]

Ideals and Pragmatism

I tend to think of methods like participatory design or the Afro-centric paradigm of design that I described above as ideals. Their goals of deeply connecting with the product's users to design technology that is truly useful to them are laudable and something I'd want to see more of in this world, where the mantra "move fast and break things" has been taken all too literally. They represent hyperlocal approaches, designed to be as inclusive as possible. This also means that the goals are great if the technology is meant for a small user group, such as the Herero tribe in Namibia you read about earlier or a specific group of elderly users. I love reading about these methods because they often teach me where I should question my assumptions—where there is cultural diversity that I was unaware of and that I can learn from.

There are a few things to consider with these ideal and hyperlocal approaches. For one, they don't easily scale to other communities and cultures. For global technology products, it's going to be very difficult for designers to immerse themselves in the innumerable cultural groups this world has to offer. (Though this should not be taken as an excuse to not start somewhere, especially because insights can translate to other groups.) There is also the problem that these ideal approaches are time-consuming, require lots of expertise, and there's usually no rulebook to follow on how to best apply them. Contrast this with industry. In my experience, there is a deep disconnect between the ideal ways of designing technology—the hyperlocal approaches—and the reality of industry practices, where products are often designed to reach as many people as possible and where universal interfaces are economically advantageous.[210] When practitioners create technology, it comes with immense time pressure imposed by the desire to compete and to stick to externally imposed timelines and budgets. As we found in our work with several companies, the financial pressure means that product changes to cater to other cultures are primarily being made in hindsight, when something noticeably fails for a specific market.[319] Expanding to other markets usually means getting a product localized so that it is ready for rollout to people speaking a certain language (whether they are distributed across the globe with different cultural backgrounds or not) and to comply with local regulations. The industry's view of culture is a location-based one—any adaptations that are being made are motivated by a user's location, not by deeper levels of culture. It's quite a strong contrast to the academic ideals of treating culture as a concept that varies from individual to individual. What this shows is that there's a huge translational divide between academia and industry: insights into the need for cultural adaptations, what their benefits are, and how to implement them are not well-communicated. But without a rulebook to follow, developers and designers have an easy excuse to avoid making progress on this social justice issue.

If there is such a big disconnect between ideal approaches and industry practice, what could a happy medium be? We've heard from many practitioners at larger companies that they often rely on local design and research teams to spot issues with technology products in a few countries that they define as primary markets. I think that's a step in the right direction, though critics would point out that those working in tech companies may be out of touch with others in their market. Their own experiences don't replace user research and participatory design workshops. Having these local design and research teams is also a luxury only the largest companies can afford. We have also seen design teams actively seeking published research to inform their work,[318] an approach that can allow extrapolating from hyperlocal research to a broader context. As an academic, I love seeing practitioners dive into research papers, but I also have to acknowledge that any literature review is extremely time-consuming and extracting relevant information a difficult undertaking.

I can think of several other ideas to improve the status quo: Record any design changes that were found to be necessary to adapt to a certain market. Perhaps create a design database that merges that knowledge within or even across companies. Ideally, make it publicly available. Fund more research to create and add to this knowledge, and to make it more easily accessible to industry. And so on. My intention here is not to come up with a list of solutions but rather to point out what I think is the obvious missing link: Academics and practitioners need to both put in a bit of work to learn each other's realities and beliefs. As many have argued before me, the two need to learn to better communicate with each other.[60,285] In my classes, I always try to highlight this rift between academia and industry because most of the students I teach go on to work for tech companies. The classes are an opportunity to teach them about the discussions academia is having before the students enter the corporate world. But if you ask me, companies should play a larger role in this by divesting some of their profits into cross-cultural research. In my eyes, it is their moral obligation to create culturally just technology. And as an added benefit, any efforts toward culturally just technology can also act as an insurance policy of sorts for companies to prevent them from causing harm. The more they put in, the more protected they are.

Local, Adapted, or Adaptable Technology

Much of what I talked about before had to do with either developing hyperlocal technology products or adapting existing products to other countries and cultures. Both have up- and downsides and neither is going to solve the problem that we don't have all that much knowledge for designing everyday technology that has global reach. We don't even have a good solution for a system architecture that can support culturally adaptive user interfaces, meaning that any efforts in this direction will require retroactively changing a technology's code base to allow for that flexibility—a hack that I'm sure would throw any software engineer into a deep depression.

There is a third option, one that involves providing ways for users to personalize the technology over time. Kristina Höök, a professor in Interaction Design at KTH Royal Institute of Technology in Stockholm, calls this "design for appropriation."[128] It's a way of designing social technologies such that users can actively change the meaning and use of a system. Of course, this also means that the technology needs to allow for flexible use; it needs to give users the freedom to shape it to suit their individual and group practices and norms. Describing a similar idea in the context of mobile phone usage, my colleague Huatong Sun refers to this consideration as "user localization:"[260] Because technology continues to be shaped and localized even after it hits the market, Sun argues that there should be a "cyclical, open-ended design process," in which designers continue to observe how the technology is being used in order to draft a more suitable design. This would mean starting off with a suboptimal technology design, and seems to place the burden on the user rather than the product developers and designers. But it may also give users the much-needed flexibility that is needed to make technology suit their cultural frames. In cases where providing a culturally adapted design is only partially feasible, I think this approach deserves our attention.

There are obviously some practical issues to solve when developing technology for a global market. But solving them first requires forming an opinion on the question of whether we should offer universal, adaptive, or adaptable technology. Technology creators need to make a choice whether to provide a one-size-fits-all interface, which is often economically advantageous,[210] or one that is culturally adapted and adaptable over time. I would hope to hear a resounding "yay for culturally adaptive technology" coming out of any of my readers' mouths right now. It's what I've been implicitly and explicitly arguing for throughout this book because I see it as our moral obligation to provide technology that is equally usable for people around the globe. On its surface, culturally adaptive systems (if done well) could certainly solve the problem of a mismatch between people's values, norms, and needs, and those of the technology. But culturally adaptive systems also raise several challenging questions, even if we ignore the economic viability for a moment. Do we want everyone in their own cultural technology bubble? What do we lose if we no longer experience culture shock in technology given that it can open our minds, unlock creativity, and let us grow? And how fine-grained do the adaptations have to be, given that there is no agreement on what constitutes culture?

I don't have a complete answer to these questions, but I find it essential that we think deeply about them to avoid another set of unanticipated consequences. To think through these questions, we could, for example, draw an analogy between people and technology. I had told you earlier that aside from people, technology with all its embedded cultural values and assumptions is yet another actor in the cultural cycle. People often make a great effort to modify their behavior to the cultural norms of others, a process that Andy Molinsky, a professor at Brandeis University's International Business School, described as "cross-cultural code switching."[188] Many of us do this all the time, often to conform to dominant norms, when we interact with people from other cultures.[36] Chances are that we don't even notice it. People who are successful at code-switching are sometimes described as "culturally intelligent."[70] Some people are more culturally intelligent than others, perhaps because they have a better radar for differences, because they find it easier to change their own behaviors, or because they have had a lot of practice trying to adapt to others. But this is where I see an opportunity: Machines could very well be better at cross-cultural code-switching than humans. Whereas people often need to deliberately and effortfully adapt to another culture, machines could trigger an adaptation much more quickly. And it gets even better: unlike people, machines would not experience the emotional challenges that sometimes prevent people from code-switching, as when the adaptation conflicts with one's values. Here's where graphical user interfaces, robots, or the latest smart speakers have a distinct advantage over us: They don't experience emotions. If developers were willing to create them, machines could come equipped with the knowledge of cultural differences, detect it in another person, and adapt accordingly—all without making any effort or experiencing psychological toll from changing their nonexisting values and without perceiving the dreaded intergroup threat. For once, technology may be better than humans.

Now, you may have just seen your alarm bells go off because a website or robot that perfectly adapts to your cultural values, norms, and preferences can certainly sound scary. After all, most of us would probably worry that a technology product would just pretend to share our values, to be in our cultural ingroup, simply to gain our trust and fool us into believing it or buying even more things that we don't need. Indeed, mimicking someone else by adjusting one's facial expressions, gestures, speech, or postures is good for building interpersonal rapport,[274] but it can also backfire when it gets consciously used as a tool to exploit someone's trust. And similarly negative effects can arise when an AI tries to adapt, as my collaborators Daniela Roesner, PhD student Jeffrey Basoah, several undergraduate students, and I observed in one of our studies.[31] When interviewing African American participants about their interaction with ChatGPT and other AI-supported writing tools, we found that the participants generally perceive ChatGPT as "White" and "not designed for us." They were reacting to the fact that ChatGPT and other AI support tools often misunderstand their dialect, African American Vernacular English, or even attempt to correct it. Sometimes

ChatGPT would try to mimic the dialect, which an African American ChatGPT user dismissed as: "[The AI is] like trying to mask and I guess imitate Black language and it's not doing it successfully." Clearly, the solution here is more complex than training an AI to become better at mimicking; any solution also needs to take into account its implications on trust and authenticity.

It's interesting, then, to think about how we can maintain a healthy middle ground between mimicry and developing technology that feels like it is designed for diverse groups of people. My hypothesis is that technology needs to become as empathetic as people are in order to find the right level of adaptation, the sweet spot. People take time to adapt to another culture and when they do, it is rarely perfect. Even for immigrants who fully immerse themselves in a new culture, it often takes years to become fluent in it. Technology design could be informed by people's cross-cultural code-switching and mimicry, both of which are slow and incomplete but effective both at retaining our experience of cultural differences, with all its up- and downsides, and at not pretending to be someone we are not.

This discussion also brings out a host of philosophical and ethical questions to grapple with, such as which cultural values one wants to preserve and whether it is morally justifiable to design against unwanted practices and values. I bet you would agree that not every cultural value is a good one; but how do we know that our moral stance is the right one? These are extremely difficult questions to ponder over, but ones that technology developers should have an answer to. A good start could be to require deep engagement with ethics, the humanities, and science and technology studies, such as by funding their university programs and adding more employees with these backgrounds.

Here's an additional step that is required to get users the culturally appropriate technology they deserve: Support the technology development in various countries and local communities. With more competition in the tech industry across the globe, we'd see more design diversity. We could learn from it and study it. (Some of this is being done already, as you've seen in this book, but I'd argue it's nowhere near enough.) As a nice side effect, we would inevitably have more discussions on the risks of exporting and importing technologies to and from companies in other countries, because the problem would suddenly affect richer countries, too. It would be an eye-opener for those whose needs, values, and preferences are closely aligned with those of the technology landscape-dominating US if they were to use more tech products designed in other countries and cultures. It could even result in new and creative design solutions that break free from design patterns that follow a "this is how it's always been done" mantra rather than nurture an ambition to find what is truly best. And, likewise, people who are currently a technology minority could see that technology can be designed in ways that feel more suitable for their cultural backgrounds.

Unfortunately, as much as I would like to, I cannot change the fact that governments and tech companies are usually uninterested in sharing their wealth. After all, this is what got us to a handful of countries fighting for the technology hegemony. But there are signs that there is more and more pushback to this power imbalance, starting with the increasing number of tech hubs around the world, as I mentioned in the introduction to this book, all the way to research coalitions that were formed to promote local technologies. One example is Masakhane, the isiZulu word for "we build together," roughly translated. Masakhane is a grassroots organization "whose mission is to strengthen and spur NLP research in African languages, for Africans, by Africans."[181] Current researchers in the Masakhane community speak at least 50 different African languages, which allows them to work on making technologies available in these languages—think machine translation all the way to conversational AI that is not just available to the small percentage of English speakers in Africa, but also to the large percentage of those who speak an African language. Masakhane supports self-organized co-creation of new technologies, which gives some of the power back to the people. I'm convinced that the cultural diversity found in this group will ultimately result in more creative, new design solutions. And the approach may well be one of the most pragmatic solutions to creating culturally just technology that is about to become infrastructure.

11

Epilogue

Sometimes, ChatGPT and I agree. For example, that time in a research meeting when a student challenged me to come up with reasons for why we should retain the cultural diversity in this world. He asked this more as a rhetorical question, right after I had explained that technology is never culturally neutral and that I believe it can spur cultural homogenization, not just by facilitating cross-cultural communication but also by simply imposing its embedded values and norms on us. The student's question is excellent to contemplate because it challenges us to fast-forward to a future in which digital technologies—if they continue being produced by only a tiny percentage of people in the world—might have converted us all to adopting similar ideas. Perhaps technologies that enhance an individual's productivity will become more highly regarded than those that foster social interaction and relationship-building. Cultural rituals or diverse communication norms may be erased by automated content moderation. Anything too colorful and playful might give way to simplistic designs. And so on.

I would not want to live in a world in which I couldn't learn about other cultures—and as a reader of this book, you might not, either. But why? After the student left my office, I asked ChatGPT about its "thoughts" on this topic. Figure 11.1 shows a part of the resulting conversation.

You

What do you think would be lost if digital technologies lead to cultural homogenization?

ChatGPT

If digital technologies were to significantly contribute to cultural homogenization, several significant aspects might be at risk of being lost:

1. **Cultural Authenticity:** Each culture has its own authentic expressions, whether in art, music, literature, or traditions. Homogenization might dilute or standardize these unique expressions, leading to a loss of authenticity.
2. **Local Knowledge and Wisdom:** Indigenous or local knowledge about agriculture, medicinal plants, sustainable practices, and traditional craftsmanship could diminish as cultures assimilate into a more homogenized global culture.
3. **Language Diversity:** Many languages are endangered or at risk of extinction. Cultural homogenization might accelerate this loss by favoring dominant languages in digital communication, leading to a decline in linguistic diversity.
4. **Cultural Practices and Rituals:** Traditional ceremonies, rituals, and practices that hold deep cultural significance might fade away or lose their relevance as cultures merge.
5. **Diverse Perspectives and Values:** Different cultures often have distinct moral values, ethical beliefs, and ways of understanding the world. Homogenization might lead to the dominance of certain perspectives while marginalizing or erasing others.
6. **Local Economies and Industries:** Globalization through digital platforms can often favor larger corporations and industries, potentially overshadowing local businesses and traditional crafts, leading to economic shifts and losses in cultural production.

Preserving cultural diversity in the face of technological advancements requires a balance between embracing the benefits of technology while also valuing and safeguarding the unique aspects of individual cultures. Encouraging digital platforms that respect and promote cultural diversity, fostering education about various cultures, and supporting initiatives that empower and preserve indigenous knowledge are essential steps in safeguarding against potential losses due to cultural homogenization.

FIGURE 11.1. A conversation with ChatGPT on the importance of cultural diversity.

It's obvious that ChatGPT is not, nor am I, a big philosopher. But we agree on the benefits of cultural diversity: Cultural diversity, even though it can be challenging in light of culture shock and cultural conflicts, is incredibly beneficial to individuals, nations, and the world. Experiencing various cultures can teach us different perspectives, ways of being, and ways of interacting with each other. It can give us new ideas, ultimately leading to more creative solutions. When we meet people with different cultural backgrounds, we also become more tolerant and open-minded, and learn about different, and often better, ways of living. Research has even shown that this gives us a chance to develop our identity, experience personal growth, and increase our self-esteem and life satisfaction.[298,322] In my eyes, these are strong arguments for more cultural diversity in technology, but note how they mostly focus on personal gain. There's another side to it, an ethical component. As ChatGPT wrote: "Homogenization might lead to the dominance of certain perspectives while marginalizing or erasing others." Not only would we lose the personal benefits that come with cultural diversity in this world, but we would also willingly accept that the majority's values may override others'. And those who have the power to do so—currently it is the US and Chinese tech industries that are most feverishly investing in a tech hegemony—seem to believe that their ways of doing things, their viewpoints, norms, and preferences, are those that everyone should have. There is an enormous responsibility in deciding what is best for the world, and it certainly necessitates a level of arrogance to believe one does.

I wrote this book to highlight the world's cultural diversity. Technology is commonly built on a very well-known set of psychological tendencies, but these tendencies primarily hold true for WEIRD people. Now that you have read about many examples of cultural diversity and tested your own cultural background in several LabintheWild studies, I hope you agree that this is an issue.

I also hope that you have become aware of common assumptions built into many technology designs. If you are yearning for a quick summary, here is a list of the top ten wrong assumptions that I think most commonly lead to digital culture shock (and, yes, many of these are WEIRD):

Misassumption 1: Technology can be transplanted from one cultural context to another. In the introduction and Chapter 2 of this book, I told you about various technology products that would very likely fail if exported to other countries and cultures, from self-driving cars that are too polite to work in places with less predictable driving behavior to apps that let people find a second wife. Behaviors, values, and norms differ across cultures, which means that technology products that invariably incorporate certain assumptions are not going to be universally suitable across the world.

Misassumption 2: People use technology with their own goals in mind. In Chapters 2 and 4, I talked about how WEIRD people are often motivated by setting their own goals in the latest fitness app or in MOOCs, prioritize their own time when scheduling events, and prefer to rely on themselves rather than on other people or on an AI when making decisions. Designs that embed this assumption often work just fine for people with an independent self-construal. But they can clash with people with an interdependent self-construal, who are more prone to being motivated by the goals of an ingroup, such as their family, friends, or co-workers.

Misassumption 3: People process information the same way. To the contrary, Chapter 3 showed how people vary in where they place most attention and how they take in and interpret visual information. This is because people's cultural experiences shape whether they are more prone to taxonomically grouping elements (an analytic thinking style, dominant in Western cultures) or whether they consider the relationship of various objects to another (as holistic thinkers tend to do). Importantly, forcing people to decipher information in a way that opposes their thinking style, such as by presenting East Asians with Western-designed web pages, or asking people to use a reading direction that is different from the one they are used to, can result in a subtle form of culture shock: a neural cost that causes people to expend extra cognitive effort and can ultimately lead them to taking longer to complete a task.

Misassumption 4: Designing for efficiency is more important than supporting social interactions. Chapter 4 described how many Western cultures have a fast pace of life and are more likely on clock time— both indicators that efficiency is of utmost importance. But many people in this world would agree that at least equally important are social interactions. Collectivists and people who tend to live on event time place a much higher emphasis on relationship-building. They may have a lower relational mobility than individualists, which means that building strong relationships is often preferred over having loose connections. For those who identify with this, technology designs that do not support interpersonal exchanges or a personal touch can be a culture shock.

Misassumption 5: People enjoy lots of choices and rely on their own decisions. In Chapters 4 and 7, you also learned that having a myriad of choices available in games, in MOOCs, or on the Web in general is not always desirable. For collectivists, choices made by members of an ingroup are often preferred and sought out before making decisions. Likewise, a large number of choices can negatively affect motivation and engagement, as shown in the case of East Asian students playing a computer game. It is therefore a bit of a WEIRD idea that people enjoy choices, and that choosing among a large number of options can even boost motivation, as it was found for individualists.

Misassumption 6: People need to be able to convey their individuality. In Chapter 5, you learned how many online social networking platforms are built around the idea that sharing personal information with a large audience is desirable. This works well for highly individualistic people who tend to use social media to create a unique identity of themselves, but it clashes with the cultural norms held by collectivists. In fact, people who tend to think of themselves more as an interdependent part of a larger group are less likely to feel the urge to set themselves apart from others. They are more concerned with upholding a collective identity and ensuring that one fits in and honors their family or greater ingroup. Hence, oversharing may be more of a concern if it means that the family or greater ingroup might be harmed.

Misassumption 7: Societies are relatively flat. Chapters 5 and 7 showed how different degrees of power distance in a society can dictate whether befriending others in online social networks follows strict social hierarchies, what level of formality is appropriate when addressing people, whether calling users by their first name is acceptable, and who gets to make decisions. Technology that communicates informally and assumes equal participation and decision-making can be a culture shock for people used to strong societal hierarchies.

Misassumption 8: Communication needs to be direct and to the point. Chapter 7 introduced the idea that much of today's technology uses and supports low-context communication, but that this can be a misfit for high-context communicators who often rely on indirect and nonverbal communication signals, such as silences, gestures, or facial expressions, when conveying and interpreting information. You have seen how high-context communicators are more likely to follow a robot's advice if it provides indirect recommendations, and vice versa for low-context communicators. I have also explained how being exposed to a different communication style than one's own can lead to frustrations and misunderstandings, a classical example of culture shock.

Misassumption 9: Noncontact cultures are the norm. In fact, contact cultures, where hugging, touching, and close proximity play an important role in building and maintaining relationships, are common in many parts of the world, such as in Arab, Latin American, and Southern European societies. Chapter 7 also described how people apply the same culturally learned norms and expectations when interacting with robots; consequently, when technology violates these norms, they can experience culture shock.

Misassumption 10: People share similar values around religion, equality, or democracy. In Chapter 8, I showed how the values upheld in various parts of the world are instead incredibly diverse and how they can clash with those expressed by technology products such as chatbots.

Throughout this book I have argued that technology designs that follow these assumptions are inadequate for their culturally diverse users. They incorporate cultural viewpoints only of their creators, who most commonly are WEIRD. And here I disagree with ChatGPT, which blatantly argued: "It's crucial to acknowledge that technology itself is neutral—it's the way it's used that can either empower or oppress cultures." No, ChatGPT. You are not neutral and neither are any of the digital technologies we encounter.

Because technology is never culturally neutral, I hope you have seen that developing it comes with a huge amount of responsibility. Digital culture shock is a powerful force triggered by seemingly innocent design decisions. It's our responsibility to scrutinize these decisions before unleashing technology into this world. But, luckily, doing so is an enormous opportunity. Just as culture shock can open our eyes to new experiences, technology can be so much more if we try to see it through a different cultural lens. Cultural diversity is an untapped source for spurring our creativity into rethinking user interactions and experiences. We should strive to understand it better.

BIBLIOGRAPHY

[1] N. Abokhodair, A. Elmadany, and W. Magdy. Holy tweets: Exploring the sharing of the Quran on Twitter. *Proceedings of the ACM on Human-Computer Interaction*, 4(CSCW2):1–32, 2020.

[2] N. Abokhodair, A. Hodges, and S. Vieweg. Photo sharing in the Arab Gulf: Expressing the collective and autonomous selves. In *Proceedings of the 2017 ACM Conference on Computer Supported Cooperative Work and Social Computing*, pages 696–711, 2017.

[3] L. Abu-Lughod. Writing against culture. In *Recapturing Anthropology: Working in the Present*, edited by Richard G. Fox, pages 137–162. Santa Fe: School of American Research, 1991.

[4] A. Acar, A. Nishimuta, D. Takamuea, K. Sakamoto, and Y. Muraki. Qualitative analysis of Facebook quitters in Japan. In *Proceedings of the Eighth International Conference on eLearning for Knowledge-Based Society*, pages 23–24. Citeseer, 2012.

[5] Accessibility.com. Complete report: 2020 website accessibility lawsuit recap, 2021. https://www.accessibility.com/complete-report-2020-website-accessibility-lawsuits.

[6] T. Adam. Digital neocolonialism and massive open online courses (MOOCs): Colonial pasts and neoliberal futures. *Learning, Media and Technology*, 44(3):365–380, 2019.

[7] U. Agarwal. The Uber Bangladesh adventure!, 2019. https://medium.com/@utsavagarwal/the-uber-bangladesh-adventure-876283b85e2.

[8] I. Ahmed. Personal communication via videoconferencing, 2023. May 12, 2023.

[9] B. Al-Ani, E. Trainer, D. Redmiles, and E. Simmons. Trust and surprise in distributed teams: Towards an understanding of expectations and adaptations. In *Proceedings of the 4th International Conference on Intercultural Collaboration*, pages 97–106, 2012.

[10] M. Ali. Towards a decolonial computing. International Society of Ethics and Information Technology, 2014.

[11] S. M. Ali. A brief introduction to decolonial computing. *XRDS: Crossroads, The ACM Magazine for Students*, 22(4):16–21, 2016.

[12] A. B. Alok, H. Sakib, S. A. Ullah, F. Huq, R. Ghosh, J. J. Mondal, M.S.I. Sakif, and J. Noor. "Khep": Exploring factors that influence the preference of contractual rides to ride-sharing apps in Bangladesh. In *Proceedings of the 6th ACM SIGCAS/SIGCHI Conference on Computing and Sustainable Societies*, pages 43–53, 2023.

[13] A. Alter. *Drunk Tank Pink: And Other Unexpected Forces that Shape How We Think, Feel, and Behave*. Penguin, 2014.

[14] https://www.amazon.com/gp/navigation-country/select-country.

[15] M. G. Ames. *The Charisma Machine: The Life, Death, and Legacy of One Laptop per Child*. MIT Press, 2019.

[16] A. Appadurai. *Modernity At Large: Cultural Dimensions of Globalization*, volume 1. University of Minnesota Press, 1996.

[17] Apple. Bring expression to your app with Genmoji, 2024. https://developer.apple.com/videos/play/wwdc2024/10220/.

[18] M. Argyle. Intercultural communication. *Cultures in Contact: Studies in Cross-Cultural Interaction*, pages 61–80, 1982.

[19] J. J. Arnett. The neglected 95%: Why American psychology needs to become less American. *American Psychological Association*, 63(7):602–614, 2008.

[20] S. A. Asongu and J. C. Nwachukwu. The mobile phone in the diffusion of knowledge for institutional quality in sub-Saharan Africa. *World Development*, 86:133–147, 2016.

[21] E. Awad, S. Dsouza, R. Kim, J. Schulz, J. Henrich, A. Shariff, J.-F. Bonnefon, and I. Rahwan. The moral machine experiment. *Nature*, 563(7729):59–64, 2018.

[22] Axim Collaborative. About the Open edX project, 2024. https://openedx.org/about-open-edx/.

[23] E. Bakshy, S. Messing, and L. A. Adamic. Exposure to ideologically diverse news and opinion on Facebook. *Science*, 348(6239):1130–1132, 2015.

[24] P. Bao, B. Hecht, S. Carton, M. Quaderi, M. Horn, and D. Gergle. Omnipedia: Bridging the Wikipedia language gap. In *Proceedings of the SIGCHI Conference on Human Factors in Computing Systems*, pages 1075–1084, 2012.

[25] W. Barber and A. Badre. Culturability: The merging of culture and usability. In *Proceedings of the 4th Conference on Human Factors and the Web*, volume 7, pages 1–10, 1998.

[26] D. Barboza. The rise of Baidu (that's Chinese for Google), September 2006. https://www.nytimes.com/2006/09/17/business/yourmoney/17baidu.html.

[27] L. M. Barna. How culture shock affects communication. *Communication*, 5(1), 1976.

[28] S. Barocas, K. Crawford, A. Shapiro, and H. Wallach. The problem with bias: Allocative versus representational harms in machine learning. In *9th Annual Conference of the Special Interest Group for Computing, Information and Society*, 2017.

[29] L. F. Barrett. *How Emotions Are Made: The Secret Life of the Brain*. Pan Macmillan, 2017.

[30] L. Baruh, E. Secinti, and Z. Cemalcilar. Online privacy concerns and privacy management: A meta-analytical review. *Journal of Communication*, 67(1):26–53, 2017.

[31] J. Basoah, J. L. Cunningham, E. Adams, A. Bose, A. Jain, K. Yadav, Z. Yang, K. Reinecke, and D. Rosner. Understanding Black users' perceptions of AI-supported writing technology. *Proceedings of the ACM on Human-Computer Interaction (CSCW2)*, 2025.

[32] K. Battarbee, J. F. Suri, and S. G. Howard. Empathy on the edge: Scaling and sustaining a human-centered approach in the evolving practice of design. IDEO. http://www.ideo.com/images/uploads/news/pdfs/Empathy_on_the_Edge.pdf, 2014.

[33] A. Baughan, T. August, N. Yamashita, and K. Reinecke. Keep it simple: How visual complexity and preferences impact search efficiency on websites. In *Proceedings of the 2020 CHI Conference on Human Factors in Computing Systems*, pages 1–10, 2020.

[34] A. Baughan, N. Oliveira, T. August, N. Yamashita, and K. Reinecke. Do cross-cultural differences in visual attention patterns affect search efficiency on websites? In *Proceedings of the 2021 CHI Conference on Human Factors in Computing Systems*, pages 1–12, 2021.

[35] S. Benartzi. *The Smarter Screen: Surprising Ways to Influence and Improve Online Behavior*. Penguin, 2017.

[36] R. Benjamin. *Race after Technology: Abolitionist Tools for the New Jim Code*. John Wiley & Sons, 2019.

[37] J. W. Berry. Stress perspectives on acculturation. *The Cambridge Handbook of Acculturation Psychology*, 1, pages 43–56, 2006.

[38] H. Beyer and K. Holtzblatt. Contextual design. *Interactions*, 6(1):32–42, 1999.

[39] N. J. Bidwell. Moving the centre to design social media in rural Africa. *AI & Society*, 31:51–77, 2016.

[40] N. J. Bidwell, S. Robinson, E. Vartiainen, M. Jones, M. J. Siya, T. Reitmaier, G. Marsden, and M. Lalmas. Designing social media for community information sharing in rural South Africa. In *Proceedings of the Southern African Institute for Computer Scientist and Information Technologists Annual Conference 2014, SAICSIT 2014: Empowered by Technology*, pages 104–114, 2014.

[41] L. Boroditsky. Metaphoric structuring: Understanding time through spatial metaphors. *Cognition*, 75(1):1–28, 2000.

[42] L. Boroditsky and A. Gaby. Remembrances of times east: Absolute spatial representations of time in an Australian aboriginal community. *Psychological Science*, 21(11):1635–1639, 2010.

[43] M. Boukes and R. Vliegenthart. News consumption and its unpleasant side effect: Studying the effect of hard and soft news exposure on mental well-being over time. *Journal of Media Psychology: Theories, Methods, and Applications*, 29(3):137–147, 2017.

[44] L. Breslow, D. E. Pritchard, J. DeBoer, G. S. Stump, A. D. Ho, and D. T. Seaton. Studying learning in the worldwide classroom research into edX's first MOOC. *Research & Practice in Assessment*, 8:13–25, 2013.

[45] D. E. Broockman, G. Ferenstein, and N. Malhotra. Predispositions and the political behavior of American economic elites: Evidence from technology entrepreneurs. *American Journal of Political Science*, 63(1):212–233, 2019.

[46] C. Brumann. Writing for culture: Why a successful concept should not be discarded. *Current Anthropology*, 40(S1):S1–S27, 1999.

[47] D. Byrne. Attitudes and attraction. In *Advances in Experimental Social Psychology*, volume 4, pages 35–89. Elsevier, 1969.

[48] B. Byung-yeul. Naver suffers shrinking online search market share, May 2023. https://www.koreatimes.co.kr/www/tech/2023/07/129_351990.html.

[49] C. L. Caldwell-Harris and A. Aycicegi. When personality and culture clash: The psychological distress of allocentrics in an individualist culture and idiocentrics in a collectivist culture. *Transcultural Psychiatry*, 43(3):331–361, 2006.

[50] E. R. Carrotte, A. M. Vella, and M. S. Lim. Predictors of "liking" three types of health and fitness-related content on social media: A cross-sectional study. *Journal of Medical Internet Research*, 17(8):e205, 2015.

[51] CB Insights. Global tech hubs report, 2018. https://www.slideshare.net/oqpied/cbinsights-global-tech-hubs-report-2018.

[52] P. Y. Chau, M. Cole, A. P. Massey, M. Montoya-Weiss, and R. M. O'Keefe. Cultural differences in the online behavior of consumers. *Communications of the ACM*, 45(10):138–143, 2002.

[53] J. Chauhan and A. Goel. An overview of MOOC in India. *International Journal of Computer Trends and Technology*, 49(2):111–120, 2017.

[54] S. Chen, X. Chen, Z. Lu, and Y. Huang. "My culture, my people, my hometown": Chinese ethnic minorities seeking cultural sustainability by video blogging. *Proceedings of the ACM on Human-Computer Interaction*, 7(CSCW1):1–30, 2023.

[55] J. Y. Chiao and B. K. Cheon. The weirdest brains in the world. *Behavioral and Brain Sciences*, 33(2-3):88–90, 2010.

[56] H. Cho, B. Knijnenburg, A. Kobsa, and Y. Li. Collective privacy management in social media: A cross-cultural validation. *ACM Transactions on Computer-Human Interaction (TOCHI)*, 25(3):1–33, 2018.

[57] H. F. Chua, J. E. Boland, and R. E. Nisbett. Cultural variation in eye movements during scene perception. *Proceedings of the National Academy of Sciences*, 102(35):12629–12633, 2005.

[58] M. Clark. The engineer who claimed a Google AI is sentient has been fired, 2022. https://www.theverge.com/2022/7/22/23274958/google-ai-engineer-blake-lemoine-chatbot-lamda-2-sentience.

[59] https://www.coca-cola.com/country-selector.

[60] L. Colusso, R. Jones, S. A. Munson, and G. Hsieh. A translational science model for HCI. In *Proceedings of the 2019 CHI Conference on Human Factors in Computing Systems*, pages 1–13, 2019.

[61] D. Cyr, C. Bonanni, J. Bowes, and J. Ilsever. Beyond trust: Web site design preferences across cultures. *Journal of Global Information Management (JGIM)*, 13(4):25–54, 2005.

[62] D. Cyr, M. Head, and H. Larios. Colour appeal in website design within and across cultures: A multi-method evaluation. *International Journal of Human-Computer Studies*, 68(1-2):1–21, 2010.

[63] A. D'Angelo. Allowing everyone to contribute to Quora, April 2021. https://quorablog.quora.com/Allowing-everyone-to-contribute-to-Quora.

[64] J. M. Darley and P. H. Gross. A hypothesis-confirming bias in labeling effects. *Journal of Personality and Social Psychology*, 44(1):20, 1983.

[65] D. Das. Personal email communication, 2023. July 12, 2023.

[66] D. Das, C. Østerlund, and B. Semaan. "Jol" or "pani"?: How does governance shape a platform's identity? *Proceedings of the ACM on Human-Computer Interaction*, 5(CSCW2):1–25, 2021.

[67] J. Devon. Accessibility awareness is on the rise, but is it turning into action? December 2021. https://techcrunch.com/2021/12/26/accessibility-awareness-is-on-the-rise-but-is-it-turning-into-action/.

[68] Y. Dong and K.-P. Lee. A cross-cultural comparative study of users' perceptions of a web page: With a focus on the cognitive styles of Chinese, Koreans and Americans. *International Journal of Design*, 2(2):19–30, 2008.

[69] P. Dourish, C. Graham, D. Randall, and M. Rouncefield. Theme issue on social interaction and mundane technologies. *Personal and Ubiquitous Computing*, 14:171–180, 2010.

[70] P. C. Earley and S. Ang. *Cultural Intelligence: Individual Interactions across Cultures*. Stanford University Press, 2003.

[71] S. Elsayed-Ali, E. Bonsignore, and J. Chan. Exploring challenges to inclusion in participatory design from the perspectives of Global North practitioners. *Proceedings of the ACM on Human-Computer Interaction*, 7(CSCW1):1–25, 2023.

[72] https://en.wikipedia.org/wiki/Bengalis.

[73] H. M. Eraqi and I. Sobh. Autonomous driving in the face of unconventional odds. *Communications of the ACM*, 64(4):64–66, 2021.

[74] G. Eresha, M. Häring, B. Endrass, E. André, and M. Obaid. Investigating the influence of culture on proxemic behaviors for humanoid robots. In *2013 IEEE Ro-Man*, pages 430–435. IEEE, 2013.

[75] H. I. Evans, M. Wong-Villacres, D. Castro, E. Gilbert, R. I. Arriaga, M. Dye, and A. Bruckman. Facebook in Venezuela: Understanding solidarity economies in low-trust environments. In *Proceedings of the 2018 CHI Conference on Human Factors in Computing Systems*, pages 1–12, 2018.

[76] V. Evers and P. Hinds. The truth about universal design: How knowledge on basic human functioning, used to inform design, differs across cultures. In *Proceedings of the 9th International Workshop on Internationalisation of Products and Systems*, pages 37–47. 2010.

[77] A. Faiola and K. F. Macdorman. The influence of holistic and analytic cognitive styles on online information design: Toward a communication theory of cultural cognitive design. *Information, Community & Society*, 11(3):348–374, 2008.

[78] A. Faiola and S. A. Matei. Cultural cognitive style and web design: Beyond a behavioral inquiry into computer-mediated communication. *Journal of Computer-Mediated Communication*, 11(1):375–394, 2005.

[79] A. Fan, H. Shen, L. Wu, A. S. Mattila, and A. Bilgihan. Whom do we trust? Cultural differences in consumer responses to online recommendations. *International Journal of Contemporary Hospitality Management*, 30(3):1508–1525, 2018.

[80] J. Fan and A. X. Zhang. Digital juries: A civics-oriented approach to platform governance. In *Proceedings of the 2020 CHI Conference on Human Factors in Computing Systems*, pages 1–14, 2020.

[81] S. R. Fitzsimmons, Y. Liao, and D. C. Thomas. From crossing cultures to straddling them: An empirical examination of outcomes for multicultural employees. *Journal of International Business Studies*, 48:63–89, 2017.

[82] J. A. Fodor. *The Language of Thought*, volume 5. Harvard University Press, 1975.

[83] B. Fogg and D. Iizawa. Online persuasion in Facebook and Mixi: A cross-cultural comparison. In *Persuasive Technology: Third International Conference, PERSUASIVE 2008, Oulu, Finland, June 4-6, 2008. Proceedings 3*, pages 35–46. Springer, 2008.

[84] N. Ford and S. Y. Chen. Matching/mismatching revisited: An empirical study of learning and teaching styles. *British Journal of Educational Technology*, 32(1):5–22, 2001.

[85] B. Friedman. Value-sensitive design. *Interactions*, 3(6):16–23, 1996.

[86] B. Friedman and D. G. Hendry. *Value Sensitive Design: Shaping Technology with Moral Imagination*. MIT Press, 2019.

[87] B. Friedman, P. Kahn, and A. Borning. Value sensitive design: Theory and methods. *University of Washington Technical Report*, 2(8), 2002.

[88] O. Fuhrman and L. Boroditsky. Cross-cultural differences in mental representations of time: Evidence from an implicit nonlinguistic task. *Cognitive Science*, 34(8):1430–1451, 2010.

[89] A. Furnham. Communicating in foreign lands: The cause, consequences and cures of culture shock. *Language, Culture and Curriculum*, 6(1):91–109, 1993.

[90] A. Furnham et al. Culture shock: A review of the literature for practitioners. *Psychology*, 10(13):1832, 2019.

[91] A. Gaby. The Thaayorre think of time like they talk of space. *Frontiers in Psychology*, 3:300, 2012.

[92] A. R. Gaby. A grammar of Kuuk Thaayorre (Vol. 74). Walter De Gruyter GmbH & Co KG. 2017.

[93] R. Gandhi, R. Veeraraghavan, K. Toyama, and V. Ramprasad. Digital green: Participatory video for agricultural extension. In *2007 International Conference on Information and Communication Technologies and Development*, pages 1–10. IEEE, 2007.

[94] B. Ganguly, S. B. Dash, D. Cyr, and M. Head. The effects of website design on purchase intention in online shopping: The mediating role of trust and the moderating role of culture. *International Journal of Electronic Business*, 8(4-5):302–330, 2010.

[95] G. Gao, S. Y. Hwang, G. Culbertson, S. R. Fussell, and M. F. Jung. Beyond information content: The effects of culture on affective grounding in instant messaging conversations. *Proceedings of the ACM on Human-Computer Interaction*, 1(CSCW):1–18, 2017.

[96] M. Gelfand. *Rule Makers, Rule Breakers: Tight and Loose Cultures and the Secret Signals that Direct Our Lives*. Scribner, 2019.

[97] M. J. Gelfand, J. L. Raver, L. Nishii, L. M. Leslie, J. Lun, B. C. Lim, L. Duan, A. Almaliach, S. Ang, J. Arnadottir, et al. Differences between tight and loose cultures: A 33-nation study. *Science*, 332(6033):1100–1104, 2011.

[98] J. Gertner. Wikipedia's moment of truth, July 2023. https://www.nytimes.com/2023/07/18/magazine/wikipedia-ai-chatgpt.html.

[99] N. Gillespie, S. Lockey, C. Curtis, J. Pool, and A. Akbari. Trust in artificial intelligence: A global study. The University of Queensland and KPMG Australia, page 10, 2023.

[100] G. Gondwe. Online incivility, hate speech and political violence in Zambia: Examining the role of online political campaign messages. *Journal of African Media Studies*, 13(1):35–51, 2021.

[101] G. Gondwe. ChatGPT and the Global South: How are journalists in sub-Saharan Africa engaging with generative AI? *Online Media and Global Communication*, 2(2):228–249, 2023.

[102] K. Goodrich and M. De Mooij. How "social" are social media? A cross-cultural comparison of online and offline purchase decision influences. *Journal of Marketing Communications*, 20(1-2):103–116, 2014.

[103] E. W. Gould, N. Zalcaria, and S. A. M. Yusof. Applying culture to web site design: A comparison of Malaysian and US web sites. In *Proceedings of the 18th Annual Conference on Computer Documentation, IEEE International Professional Communication Conference*, pages 161–171, 2000.

[104] J. L. Graham, A. T. Mintu, and W. Rodgers. Explorations of negotiation behaviors in ten foreign cultures using a model developed in the United States. *Management Science*, 40(1):72–95, 1994.

[105] M. Graham. The problem with Wikidata, April 2012. https://www.theatlantic.com/technology/archive/2012/04/the-problem-with-wikidata/255564/.

[106] W. Gravett. Digital neo-colonialism: The Chinese model of Internet sovereignty in Africa. *African Human Rights Law Journal*, 20(1):125–146, 2020.

[107] P. J. Guo and K. Reinecke. Demographic differences in how students navigate through MOOCs. In *Proceedings of the First ACM Learning @ Scale Conference*, pages 21–30, 2014.

[108] A. H. Gutchess, R. C. Welsh, A. Boduroğlu, and D. C. Park. Cultural differences in neural function associated with object processing. *Cognitive, Affective, & Behavioral Neuroscience*, 6(2):102–109, 2006.

[109] N. Guyatt. The weirdest people in the world review—A theory-of-everything study, November 2020. https://www.theguardian.com/books/2020/nov/20/the-weirdest-people-in-the-world-review-a-theory-of-everything-study.

[110] K.-F. Halamandaris and K. Power. Individual differences, dysfunctional attitudes, and social support: A study of the psychosocial adjustment to university life of home students. *Personality and Individual Differences*, 22(1):93–104, 1997.

[111] E. T. Hall. *The Hidden Dimension*, volume 609. Anchor, 1966.

[112] E. T. Hall. *The Silent Language*. Anchor, 1973.

[113] E. T. Hall. *Beyond Culture*. Anchor, 1976.

[114] M. Y. G. Hamedani and H. R. Markus. Understanding culture clashes and catalyzing change: A culture cycle approach. *Frontiers in Psychology*, 10:700, 2019.

[115] N. Hara, P. Shachaf, and K. F. Hew. Cross-cultural analysis of the Wikipedia community. *Journal of the American Society for Information Science and Technology*, 61(10):2097–2108, 2010.

[116] J. R. Harrington and M. J. Gelfand. Tightness-looseness across the 50 United States. *Proceedings of the National Academy of Sciences*, 111(22):7990–7995, 2014.

[117] Harvard. Q&A on WEIRD: A Q&A with Joseph Henrich, 2023. https://weirdpeople.fas.harvard.edu/qa-weird.

[118] D. B. Haun, C. J. Rapold, J. Call, G. Janzen, and S. C. Levinson. Cognitive cladistics and cultural override in hominid spatial cognition. *Proceedings of the National Academy of Sciences*, 103(46):17568–17573, 2006.

[119] H. A. He, N. Yamashita, C. Wacharamanotham, A. B. Horn, J. Schmid, and E. M. Huang. Two sides to every story: Mitigating intercultural conflict through automated feedback and shared self-reflections in global virtual teams. *Proceedings of the ACM on Human-Computer Interaction*, 1(CSCW):1–21, 2017.

[120] T. Hedden, S. Ketay, A. Aron, H. R. Markus, and J. D. Gabrieli. Cultural influences on neural substrates of attentional control. *Psychological Science*, 19(1):12–17, 2008.

[121] J. Henrich. *The WEIRDest People in the World: How the West Became Psychologically Peculiar and Particularly Prosperous.* Penguin UK, 2020.

[122] J. Henrich, S. J. Heine, and A. Norenzayan. The weirdest people in the world? *Behavioral and Brain Sciences*, 33(2-3):61–83, 2010.

[123] P. Hinds, L. Liu, and J. Lyon. Putting the global in global work: An intercultural lens on the practice of cross-national collaboration. *Academy of Management Annals*, 5(1):135–188, 2011.

[124] G. Hofstede. *Culture's Consequences.* Sage, 1980.

[125] G. Hofstede. Cultural differences in teaching and learning. *International Journal of Intercultural Relations*, 10(3):301–320, Jan. 1986.

[126] G. Hofstede, G. J. Hofstede, and M. Minkov. *Cultures and Organizations: Software of the Mind. Intercultural Cooperation and Its Importance for Survival.* McGraw-Hill, 2010.

[127] https://home.unicode.org/membership/members/.

[128] K. Höök. Designing familiar open surfaces. In *Proceedings of the 4th Nordic Conference on Human-Computer Interaction: Changing Roles*, pages 242–251, 2006.

[129] R. Inglehart, C. Haerpfer, A. Moreno, C. Welzel, K. Kizilova, J. Diez-Medrano, M. Lagos, P. Norris, E. Ponarin, and B. P. et al. (eds.). World values survey: All rounds, 2014. Country-Pooled Datafile Version: https://www.worldvaluessurvey.org/WVSDocumentationWVL.jsp.

[130] R. Inglehart and C. Welzel. Changing mass priorities: The link between modernization and democracy. *Perspectives on Politics*, 8(2):551–567, 2010.

[131] L. Irani, J. Vertesi, P. Dourish, K. Philip, and R. E. Grinter. Postcolonial computing: A lens on design and development. In *Proceedings of the SIGCHI Conference on Human Factors in Computing Systems*, pages 1311–1320, 2010.

[132] S. S. Iyengar and M. R. Lepper. Rethinking the value of choice: A cultural perspective on intrinsic motivation. *Journal of Personality and Social Psychology*, 76(3):349–366, 1999.

[133] L. Jane, L. Ilene, J. A. Landay, and J. R. Cauchard. Drone & wo: Cultural influences on human-drone interaction techniques. In *Proceedings of the 2017 CHI Conference on Human Factors in Computing Systems*, pages 6794–6799, 2017.

[134] L.-J. Ji, Z. Zhang, and R. E. Nisbett. Is it culture or is it language? Examination of language effects in cross-cultural research on categorization. *Journal of Personality and Social Psychology*, 87(1):57–65, 2004.

[135] M. F. Jung. Affective grounding in human-robot interaction. In *Proceedings of the 2017 ACM/IEEE International Conference on Human-Robot Interaction*, pages 263–273, 2017.

[136] J. Kasera, J. O'Neill, and N. J. Bidwell. Sociality, tempo & flow: Learning from Namibian ridesharing. In *Proceedings of the First African Conference on Human Computer Interaction*, pages 36–47, 2016.

[137] S. Kayan, S. R. Fussell, and L. D. Setlock. Cultural differences in the use of instant messaging in Asia and North America. In *Proceedings of the 2006 20th Anniversary Conference on Computer Supported Cooperative Work*, pages 525–528, 2006.

[138] I. Kayes, N. Kourtellis, D. Quercia, A. Iamnitchi, and F. Bonchi. Cultures in community question answering. In *Proceedings of the 26th ACM Conference on Hypertext & Social Media*, pages 175–184, 2015.

[139] T. Kene-Okafor. Money Fellows, an Egyptian fintech digitizing money circles, raises $31m funding, October 2022. https://techcrunch.com/2022/10/31/money-fellows -an-egyptian-fintech-digitizing-money-circles-raises-31m-funding/.

[140] J. Kersey. The Japanese search engine market—platforms, popularity and user trends, June 2023. https://www.humblebunny.com/japanese-search-engine-market -platforms-popularity-user-trends/.

[141] E. Kim. 8 successful tech leaders who overcame a learning disability called dyslexia, 2015. https://www.businessinsider.com/tech-leaders-with-dyslexia-2015-2.

[142] H. Kim and H. R. Markus. Deviance or uniqueness, harmony or conformity? A cultural analysis. *Journal of Personality and Social Psychology*, 77(4):785–800, 1999.

[143] P. Kimura-Thollander. Personal email communication, 2023. August 23, 2023.

[144] P. Kimura-Thollander and N. Kumar. Examining the "global" language of emojis: Designing for cultural representation. In *Proceedings of the 2019 CHI Conference on Human Factors in Computing Systems*, pages 1–14, 2019.

[145] S. Kitayama, S. Duffy, T. Kawamura, and J. T. Larsen. Perceiving an object and its context in different cultures: A cultural look at New Look. *Psychological Science*, 14(3):201–206, 2003.

[146] S. Kitayama, H. Park, A. T. Sevincer, M. Karasawa, and A. K. Uskul. A cultural task analysis of implicit independence: Comparing North America, Western Europe, and East Asia. *Journal of Personality and Social Psychology*, 97(2):236–255, 2009.

[147] S. Kitayama and J. Park. Cultural neuroscience of the self: Understanding the social grounding of the brain. *Social Cognitive and Affective Neuroscience*, 5(2-3):111–129, 2010.

[148] A. Kralisch, M. Eisend, and B. Berendt. The impact of culture on website navigation behaviour. In *Proc. HCI-International*, pages 1–9, 2005.

[149] M. H. Kuhn and T. S. McPartland. An empirical investigation of self-attitudes. *American Sociological Review*, 19(1):68–76, 1954.

[150] N. Kumar, N. Jafarinaimi, and M. Bin Morshed. Uber in Bangladesh: The tangled web of mobility and justice. *Proceedings of the ACM on Human-Computer Interaction*, 2(CSCW):1–21, 2018.

[151] S. Kurita. MOMA ♥ emoji, 2010. https://www.moma.org/interactives/moma _through_time/2010/acquisition-of-and-emoji/.

[152] M. Kwet. Digital colonialism: US empire and the new imperialism in the Global South. *Race & Class*, 60(4):3–26, 2019.

[153] R. N. Levine. *A Geography of Time: On Tempo, Culture, and the Pace of Life*. Basic Books, 2008.

[154] R. V. Levine and A. Norenzayan. The pace of life in 31 countries. *Journal of Cross-Cultural Psychology*, 30(2):178–205, 1999.

[155] T. Levitt. The globalization of markets. *Harvard Business Review*, 61 (May/June), 92–102. 1983.

[156] H. Li, S. Milani, V. Krishnamoorthy, M. Lewis, and K. Sycara. Perceptions of domestic robots' normative behavior across cultures. In *Proceedings of the 2019 AAAI/ACM Conference on AI, Ethics, and Society*, pages 345–351, 2019.

[157] Y. Li, A. Kobsa, B. P. Knijnenburg, and M. C. Nguyen. Cross-cultural privacy prediction. In *Proceedings on Privacy Enhancing Technologies*, pages 113–132, 2017.

[158] J. O. Liegle and T. N. Janicki. The effect of learning styles on the navigation needs of web-based learners. *Computers in Human Behavior*, 22(5):885–898, 2006.

[159] L. Y. Lin, J. E. Sidani, A. Shensa, A. Radovic, E. Miller, J. B. Colditz, B. L. Hoffman, L. M. Giles, and B. A. Primack. Association between social media use and depression among US young adults. *Depression and Anxiety*, 33(4):323–331, 2016.

[160] G. Linden, B. Smith, and J. York. Amazon.com recommendations: Item-to-item collaborative filtering. *IEEE Internet Computing*, 7(1):76–80, 2003.

[161] G. Lindgaard, C. Dudek, D. Sen, L. Sumegi, and P. Noonan. An exploration of relations between visual appeal, trustworthiness and perceived usability of homepages. *ACM Trans. Comput.-Hum. Interact.*, 18(1):1–30, Apr. 2011.

[162] G. Lindgaard, G. Fernandes, C. Dudek, and J. Brown. Attention web designers: You have 50 milliseconds to make a good first impression! *Behaviour & Information Technology*, 25(2):115–126, 2006.

[163] S. Lindtner, K. Anderson, and P. Dourish. Cultural appropriation: Information technologies as sites of transnational imagination. In *Proceedings of the ACM 2012 Conference on Computer Supported Cooperative Work*, pages 77–86, 2012.

[164] S. M. Lindtner. *Prototype Nation: China and the Contested Promise of Innovation*, volume 29. Princeton University Press, 2020.

[165] S. Linxen, C. Sturm, F. Brühlmann, V. Cassau, K. Opwis, and K. Reinecke. How weird is CHI? In *Proceedings of the 2021 CHI Conference on Human Factors in Computing Systems*, pages 1–14, 2021.

[166] S. Liu, T. Liang, S. Shao, and J. Kong. Evaluating localized MOOCs: The role of culture on interface design and user experience. *IEEE Access*, 8:107927–107940, 2020.

[167] K. D. Lo and C. Houkamau. Exploring the cultural origins of differences in time orientation between European New Zealanders and Māori. *NZJHRM*, 12(3):105–123, 2012.

[168] X. Lu, W. Ai, X. Liu, Q. Li, N. Wang, G. Huang, and Q. Mei. Learning from the ubiquitous language: An empirical analysis of emoji usage of smartphone users. In *Proceedings of the 2016 ACM International Joint Conference on Pervasive and Ubiquitous Computing*, pages 770–780, 2016.

[169] E. Luger and A. Sellen. "Like having a really bad PA": The gulf between user expectation and experience of conversational agents. In *Proceedings of the 2016 CHI Conference on Human Factors in Computing Systems*, pages 5286–5297, 2016.

[170] S. Lysgaand. Adjustment in a foreign society: Norwegian Fulbright grantees visiting the United States. *International Social Science Bulletin*, 1955.

[171] V. Ma and T. J. Schoeneman. Individualism versus collectivism: A comparison of Kenyan and American self-concepts. *Basic and Applied Social Psychology*, 19(2):261–273, 1997.

[172] I. MacKenzie, C. Meyer, and S. Noble. How retailers can keep up with consumers. *McKinsey & Company*, 18(1), 2013.

[173] E. A. Maguire, D. G. Gadian, I. S. Johnsrude, C. D. Good, J. Ashburner, R. S. Frackowiak, and C. D. Frith. Navigation-related structural change in the hippocampi of taxi drivers. *Proceedings of the National Academy of Sciences*, 97(8):4398–4403, 2000.

[174] A. Majid, M. Bowerman, S. Kita, D. B. Haun, and S. C. Levinson. Can language restructure cognition? The case for space. *Trends in Cognitive Sciences*, 8(3):108–114, 2004.

[175] L. Mann, M. Radford, P. Burnett, S. Ford, M. Bond, K. Leung, H. Nakamura, G. Vaughan, and K.-S. Yang. Cross-cultural differences in self-reported decision-making style and confidence. *International Journal of Psychology*, 33(5):325–335, 1998.

[176] A. Marcus and E. W. Gould. Crosscurrents: Cultural dimensions and global web user-interface design. *Interactions*, 7(4):32–46, 2000.

[177] P. Marentette, P. Pettenati, A. Bello, and V. Volterra. Gesture and symbolic representation in Italian and English-speaking Canadian 2-year-olds. *Child Development*, 87(3):944–961, 2016.

[178] H. R. Markus and S. Kitayama. Culture and the self: Implications for cognition, emotion, and motivation. *Psychological Review*, 98(2):224–253, 1991.

[179] H. R. Markus and S. Kitayama. Cultures and selves: A cycle of mutual constitution. *Perspectives on Psychological Science*, 5(4):420–430, 2010.

[180] H. R. Markus and B. Schwartz. Does choice mean freedom and well-being? *Journal of Consumer Research*, 37(2):344–355, 2010.

[181] https://www.masakhane.io/.

[182] T. Masuda and R. E. Nisbett. Attending holistically versus analytically: Comparing the context sensitivity of Japanese and Americans. *Journal of Personality and Social Psychology*, 81(5):922–934, 2001.

[183] B. McSweeney. Hofstede's model of national cultural differences and their consequences: A triumph of faith—a failure of analysis. *Human Relations*, 55(1):89–118, 2002.

[184] K. Mei, N. Oliveira, R. Barragan, Y. Tsvetkov, A. Meltzoff, M. Sap, and K. Reinecke. How conversational AI triggers symptoms similar to culture shock. In preparation.

[185] D. Metaxa-Kakavouli, K. Wang, J. A. Landay, and J. Hancock. Gender-inclusive design: Sense of belonging and bias in web interfaces. In *Proceedings of the 2018 CHI Conference on Human Factors in Computing Systems*, pages 1–6, 2018.

[186] Y. Miyamoto, R. E. Nisbett, and T. Masuda. Culture and the physical environment: Holistic versus analytic perceptual affordances. *Psychological Science*, 17(2):113–119, 2006.

[187] Q. Mkabela. Using the Afrocentric method in researching indigenous African culture. *The Qualitative Report*, 10(1):178–189, 2005.

[188] A. Molinsky. Cross-cultural code-switching: The psychological challenges of adapting behavior in foreign cultural interactions. *Academy of Management Review*, 32(2):622–640, 2007.

[189] F. T. Moura, N. Singh, and W. Chun. The influence of culture in website design and users' perceptions: Three systematic reviews. *Journal of Electronic Commerce Research*, 17(4):312–339, 2016.

[190] M. J. Muller and S. Kuhn. Participatory design. *Communications of the ACM*, 36(6):24–28, 1993.

[191] A. Munoriyarwa, S. Chiumbu, and G. Motsaathebe. Artificial intelligence practices in everyday news production: The case of South Africa's mainstream newsrooms. *Journalism Practice*, 17(7):1374–1392, 2023.

[192] J. Na, M. Kosinski, and D. J. Stillwell. When a new tool is introduced in different cultural contexts: Individualism-collectivism and social network on Facebook. *Journal of Cross-Cultural Psychology*, 46(3):355–370, 2015.

[193] M. Nakada and T. Tamura. Japanese conceptions of privacy: An intercultural perspective. *Ethics and Information Technology*, 7:27–36, 2005.

[194] K. K. Nam, M. S. Ackerman, and L. A. Adamic. Questions in, knowledge in? A study of Naver's question answering community. In *Proceedings of the SIGCHI Conference on Human Factors in Computing Systems*, pages 779–788, 2009.

[195] D. Nemer and J. O'Neill. Rethinking MOOCs: The promises for better education in India. *International Journal of Information Communication Technologies and Human Development (IJICTHD)*, 11(1):36–50, 2019.

[196] H. H. Nigatu, J. Canny, and S. E. Chasins. Low-resourced languages and online knowledge repositories: A need-finding study. In *Proceedings of the CHI Conference on Human Factors in Computing Systems*, pages 1–21, 2024.

[197] R. Nisbett. *The Geography of Thought: How Asians and Westerners Think Differently... and Why.* Simon and Schuster, 2004.

[198] R. E. Nisbett, K. Peng, I. Choi, and A. Norenzayan. Culture and systems of thought: Holistic versus analytic cognition. *Psychological Review*, 108(2):291–310, 2001.

[199] M. Nkwo and R. Orji. Persuasive technology in African context: Deconstructing persuasive techniques in an African online marketplace. In *Proceedings of the Second African Conference for Human Computer Interaction: Thriving Communities*, pages 1–10, 2018.

[200] M. Nkwo, R. Orji, J. C. Nwokeji, and C. Ndulue. E-commerce personalization in Africa: A comparative analysis of Jumia and Konga. In *Proceedings of the 3rd International CEUR Workshop on Personalization in Persuasive Technology*, pages 68–76, 2018.

[201] S. U. Noble. Algorithms of oppression: How search engines reinforce racism. In *Algorithms of Oppression*. New York University press, 2018.

[202] M. Nordhoff, T. August, N. A. Oliveira, and K. Reinecke. A case for design localization: Diversity of website aesthetics in 44 countries. In *Proceedings of the 2018 CHI Conference on Human Factors in Computing Systems*, pages 1–12, 2018.

[203] P. Norris and R. Inglehart. *Cosmopolitan Communications: Cultural Diversity in a Globalized World*. Cambridge University Press, 2009.

[204] P. Norris and R. F. Inglehart. Muslim integration into western cultures: Between origins and destinations. *Political Studies*, 60(2):228–251, 2012.

[205] K. Oberg. Cultural shock: Adjustment to new cultural environments. *Practical Anthropology*, (4):177–182, 1960.

[206] N. Oliveira, N. Andrade, and K. Reinecke. Participation differences in Q&A sites across countries: Opportunities for cultural adaptation. In *Proceedings of the 9th Nordic Conference on Human-Computer Interaction*, page 6. ACM, 2016.

[207] N. Oliveira, M. Muller, N. Andrade, and K. Reinecke. The exchange in StackExchange: Divergences between Stack Overflow and its culturally diverse participants. *Proceedings of the ACM on Human-Computer Interaction*, 2(CSCW):1–22, 2018.

[208] J. S. Olson and G. M. Olson. Culture surprises in remote software development teams: "When in Rome" doesn't help when your team crosses time zones, and your deadline doesn't. *Queue*, 1(9):52–59, 2003.

[209] A. S. Ong and C. Ward. The construction and validation of a social support measure for sojourners: The Index of Sojourner Social Support (ISSS) scale. *Journal of Cross-Cultural Psychology*, 36(6):637–661, 2005.

[210] R. Orji and R. L. Mandryk. Developing culturally relevant design guidelines for encouraging healthy eating behavior. *International Journal of Human-Computer Studies*, 72(2):207–223, 2014.

[211] K. Oyibo and J. Vassileva. Investigation of the moderating effect of culture on users' susceptibility to persuasive features in fitness applications. *Information*, 10(11):344, 2019.

[212] D. Oyserman, H. M. Coon, and M. Kemmelmeier. Rethinking individualism and collectivism: Evaluation of theoretical assumptions and meta-analyses. *Psychological Bulletin*, 128(1):3–72, 2002.

[213] U. S. Pawar, J. Pal, and K. Toyama. Multiple mice for computers in education in developing countries. In *2006 International Conference on Information and Communication Technologies and Development*, pages 64–71. IEEE, 2006.

[214] P. J. Pelto. The differences between "tight" and "loose" societies. *Trans-action*, 5:37–40, 1968.

[215] A. N. Peters, H. Winschiers-Theophilus, and B. E. Mennecke. Cultural influences on Facebook practices: A comparative study of college students in Namibia and the United States. *Computers in Human Behavior*, 49:259–271, 2015.

[216] J. Pflug. Contextuality and computer-mediated communication: A cross cultural comparison. *Computers in Human Behavior*, 27(1):131–137, 2011.

[217] S. Pinker. *The Language Instinct: How the Mind Creates Language*. Penguin UK, 2003.

[218] J.-C. Plantin, C. Lagoze, P. N. Edwards, and C. Sandvig. Infrastructure studies meet platform studies in the age of Google and Facebook. *New Media & Society*, 20(1):293–310, 2018.

[219] L. Praslova. Neurodivergent people make great leaders, not just employees, December 2021. https://www.fastcompany.com/90706149/neurodivergent-people-make-great -leaders-not-just-employees.

[220] M. Price, A. C. Legrand, Z. M. Brier, K. van Stolk-Cooke, K. Peck, P. S. Dodds, C. M. Danforth, and Z. W. Adams. Doomscrolling during Covid-19: The negative association between daily social and traditional media consumption and mental health symptoms during the Covid-19 pandemic. *Psychological Trauma: Theory, Research, Practice, and Policy*, 14(8):1338–1346, 2022.

[221] R. Qadri, R. Shelby, C. L. Bennett, and E. Denton. AI's regimes of representation: A community-centered study of text-to-image models in South Asia. In *Proceedings of the 2023 ACM Conference on Fairness, Accountability, and Transparency*, pages 506–517, 2023.

[222] R. Redfield, R. Linton, and M. J. Herskovits. Memorandum for the study of acculturation. *American Anthropologist*, 38(1):149–152, 1936.

[223] K. Reinecke and A. Bernstein. Tell me where you've lived, and I'll tell you what you like: Adapting interfaces to cultural preferences. In *International Conference on User Modeling, Adaptation, and Personalization (UMAP)*, pages 185–196, 2009.

[224] K. Reinecke and A. Bernstein. Improving performance, perceived usability, and aesthetics with culturally adaptive user interfaces. *ACM ToCHI*, 18(2), 2011.

[225] K. Reinecke and A. Bernstein. Knowing what a user likes: A design science approach to interfaces that automatically adapt to culture. *MIS Quarterly*, 37(2):427–453, 2013.

[226] K. Reinecke and K. Z. Gajos. Quantifying visual preferences around the world. In *Proc. CHI' 14*, pages 11–20. ACM, 2014.

[227] K. Reinecke and K. Z. Gajos. LabintheWild: Conducting large-scale online experiments with uncompensated samples. In *Proceedings of the 18th ACM Conference on Computer Supported Cooperative Work & Social Computing*, CSCW '15, pages 1364–1378, 2015.

[228] K. Reinecke, M. K. Nguyen, A. Bernstein, M. Näf, and K. Z. Gajos. Doodle around the world: Online scheduling behavior reflects cultural differences in time perception and group decision-making. In *Proceedings of the 2013 Conference on Computer Supported Cooperative Work*, pages 45–54, 2013.

[229] K. Reinecke, T. Yeh, L. Miratrix, R. Mardiko, Y. Zhao, J. Liu, and K. Gajos. Predicting users' first impressions of website aesthetics with a quantification of perceived visual complexity and colorfulness. In *Proc. CHI' 13*, pages 2049–2058. ACM, 2013.

[230] E. Reutskaja, N. N. Cheek, S. Iyengar, and B. Schwartz. Choice deprivation, choice overload, and satisfaction with choices across six nations. *Journal of International Marketing*, 30(3):18–34, 2022.

[231] D. M. Romero, K. Reinecke, and L. P. Robert Jr. The influence of early respondents: Information cascade effects in online event scheduling. In *Proceedings of the Tenth ACM International Conference on Web Search and Data Mining*, pages 101–110, 2017.

[232] R. Rosenthal and L. Jacobson. Pygmalion in the classroom. *The Urban Review*, 3(1): 16–20, 1968.

[233] N. Rowe. Underage workers are training AI, 2023. https://www.wired.com/story /artificial-intelligence-data-labeling-children/?bxid=5bd66dda2ddf9c6194382c71&cn did=9637186&esrc=MC_load&hasha=af14c87718f3d0f7c9fd9a1831fec0ef&hashb=

2db9bf8d45aaef08247e21f91d63f5a45e07c68a&source=Email_0_EDT_WIR_NEW
SLETTER_0_BACKCHANNEL_ZZ&utm_mailing=WIR_Backchannel_111723.

[234] F. Rudmin. Constructs, measurements and models of acculturation and acculturative stress. *International Journal of Intercultural Relations*, 33(2):106–123, 2009.

[235] D. Russell. Personal communication via video conference, 2023. July 24, 2023.

[236] D. Russell. Personal email communication, 2023. July 20, 2023.

[237] D. M. Russell. *The Joy of Search: A Google Insider's Guide to Going Beyond the Basics*. MIT Press, 2023.

[238] J. Russell. Uber launches in Bangladesh to continue its emerging market push, 2016. https://techcrunch.com/2016/11/22/uber-bangladesh/.

[239] R. M. Ryan and E. L. Deci. Self-determination theory and the facilitation of intrinsic motivation, social development, and well-being. *American Psychologist*, 55(1):68–78, 2000.

[240] M. Saleh, M. Khamis, and C. Sturm. What about my privacy, habibi? Understanding privacy concerns and perceptions of users from different socioeconomic groups in the Arab world. In *Human-Computer Interaction—Proceedings, INTERACT 2019: 17th IFIP TC 13 International Conference*, pages 67–87. Springer, 2019.

[241] Y. Sato. Wikipedia has a language problem. Here's how to fix it. https://undark.org/2021/08/12/wikipedia-has-a-language-problem-heres-how-to-fix-it/.

[242] K. Savani, H. R. Markus, N. Naidu, S. Kumar, and N. Berlia. What counts as a choice? US Americans are more likely than Indians to construe actions as choices. *Psychological Science*, 21(3):391–398, 2010.

[243] D. Schuler and A. Namioka. *Participatory Design: Principles and Practices*. CRC Press, 1993.

[244] K. Seaborn, G. Barbareschi, and S. Chandra. Not only WEIRD but "uncanny"? A systematic review of diversity in human-robot interaction research. *International Journal of Social Robotics*, vol. 15, pages 1841–1870, 2023.

[245] B. Semaan, B. Dosono, and L. M. Britton. Impression management in high context societies: 'Saving face' with ICT. In *Proceedings of the 2017 ACM Conference on Computer Supported Cooperative Work and Social Computing*, pages 712–725, 2017.

[246] https://www.semrush.com/website/quora.com/overview/.

[247] A. Seth, J. Cao, X. Shi, R. Dotsch, Y. Liu, and M. W. Bos. Cultural differences in friendship network behaviors: A Snapchat case study. In *Proceedings of the 2023 CHI Conference on Human Factors in Computing Systems*, pages 1–14, 2023.

[248] D. Shah. XuetangX: A look at China's first and biggest MOOC platform. https://www.classcentral.com/report/xuetangx/, year=2016, month=October.

[249] H. Shen, C. Faklaris, H. Jin, L. Dabbish, and J. I. Hong. 'I can't even buy apples if I don't use mobile pay?' When mobile payments become infrastructural in China. *Proceedings of the ACM on Human-Computer Interaction*, 4(CSCW2):1–26, 2020.

[250] S. Shen, H. Tennent, H. Claure, and M. Jung. My telepresence, my culture? An intercultural investigation of telepresence robot operators' interpersonal distance behaviors. In *Proceedings of the 2018 CHI Conference on Human Factors in Computing Systems*, pages 1–11, 2018.

[251] M. Silic, D. Cyr, A. Back, and A. Holzer. Effects of color appeal, perceived risk and culture on user's decision in presence of warning banner message. In *Proceedings of the 50th Hawaii International Conference on System Sciences*, 2017.

[252] N. Singh, O. Furrer, and M. Ostinelli. To localize or to standardize on the Web: Empirical evidence from Italy, India, Netherlands, Spain, and Switzerland. *Multinational Business Review*, 12(1):69–88, 2004.

[253] N. Singh and H. Matsuo. Measuring cultural adaptation on the Web: A content analytic study of US and Japanese web sites. *Journal of Business Research*, 57(8):864–872, 2004.

[254] N. Singh, H. Zhao, and X. Hu. Analyzing the cultural content of web sites: A cross-national comparision of China, India, Japan, and US. *International Marketing Review*, 22(2):129–146, 2005.

[255] A. Sorokowska, P. Sorokowski, P. Hilpert, K. Cantarero, T. Frackowiak, K. Ahmadi, A. M. Alghraibeh, R. Aryeetey, A. Bertoni, K. Bettache, et al. Preferred interpersonal distances: A global comparison. *Journal of Cross-Cultural Psychology*, 48(4):577–592, 2017.

[256] J. Spencer-Rodgers and T. McGovern. Attitudes toward the culturally different: The role of intercultural communication barriers, affective responses, consensual stereotypes, and perceived threat. *International Journal of Intercultural Relations*, 26(6):609–631, 2002.

[257] Star Online Report. Quora launches Bengali service, January 2019. https://www.thedailystar.net/online/quora-bangla-service-launched-successfully-1691200.

[258] W. G. Stephan and C. W. Stephan. An integrated threat theory of prejudice. In *Reducing Prejudice and Discrimination*, pages 23–45. Psychology Press, 2013.

[259] S. Sultana and S. I. Ahmed. Witchcraft and HCI: Morality, modernity, and postcolonial computing in rural Bangladesh. In *Proceedings of the 2019 CHI Conference on Human Factors in Computing Systems*, pages 1–15, 2019.

[260] H. Sun. *Cross-Cultural Technology Design: Creating Culture-Sensitive Technology for Local Users*. OUP USA, 2012.

[261] L. Sun, W. Zhan, M. Tomizuka, and A. D. Dragan. Courteous autonomous cars. In *2018 IEEE/RSJ International Conference on Intelligent Robots and Systems (IROS)*, pages 663–670. IEEE, 2018.

[262] N. M. Sussman and H. M. Rosenfeld. Influence of culture, language, and sex on conversational distance. *Journal of Personality and Social Psychology*, 42(1):66–74, 1982.

[263] M. Szymanski, S. R. Fitzsimmons, and W. M. Danis. Multicultural managers and competitive advantage: Evidence from elite football teams. *International Business Review*, 28(2):305–315, 2019.

[264] M. Szymanski and K. Kalra. Performance effects of interaction between multicultural managers and multicultural team members: Evidence from elite football competitions. *Thunderbird International Business Review*, 63(2):235–251, 2021.

[265] T. Talhelm, L. Wei, R. Sun, D. Medvedev, A. San Martin, M. Helmy, A. Samekin, A. Z. Scherman, A. S. English, and the Responsibilism Collaboration Team. Collectivism isn't what people think it is: A study of 100 cultures, 2024. Under submission.

[266] T. Talhelm, X. Zhang, S. Oishi, C. Shimin, D. Duan, X. Lan, and S. Kitayama. Large-scale psychological differences within China explained by rice versus wheat agriculture. *Science*, 344(6184):603–608, 2014.

[267] A. Tanaka. Google to expand local development teams in Asian countries, September 2022. https://asia.nikkei.com/Business/Technology/Google-to-expand-local-development-teams-in-Asian-countries.

[268] T. Thadani. California suspends cruise's autonomous driverless vehicle permits, October 2023. https://www.washingtonpost.com/technology/2023/10/24/cruise-robotaxis-california-suspended/

[269] The Culture Factor Group. Frequently asked questions, 2023. https://www.hofstede-insights.com/frequently-asked-questions.

[270] The Editors of Encyclopaedia Britannica. Bengali language, 2023. https://www.britannica.com/topic/Bengali-language.

[271] The Waymo Team. First million rider-only miles: How the Waymo driver is improving road safety, February 2023. https://waymo.com/blog/2023/02/first-million-rider-only-miles-how.html.

[272] R. Thomson, M. Yuki, T. Talhelm, J. Schug, M. Kito, A. H. Ayanian, J. C. Becker, M. Becker, C.-Y. Chiu, H.-S. Choi, et al. Relational mobility predicts social behaviors in 39 countries and is tied to historical farming and threat. *Proceedings of the National Academy of Sciences*, 115(29):7521–7526, 2018.

[273] A. de. Tocqueville. Democracy in America. In *Democracy: A Reader*, pages 67–76. Columbia University Press, 2016. Original work published 1835.

[274] C. L. Toma. Towards conceptual convergence: An examination of interpersonal adaptation. *Communication Quarterly*, 62(2):155–178, 2014.

[275] A. Tomer. America's commuting choices: 5 major takeaways from 2016 census data, October 2017. https://www.brookings.edu/articles/americans-commuting-choices-5-major-takeaways-from-2016-census-data/.

[276] K. Toyama. *Geek Heresy: Rescuing Social Change from the Cult of Technology*. Public-Affairs, 2015.

[277] N. Tractinsky, A. S. Katz, and D. Ikar. What is beautiful is usable. *Interacting with Computers*, 13(2):127–145, 2000.

[278] S. Trauzettel-Klosinski, K. Dietz, I. S. Group, et al. Standardized assessment of reading performance: The new international reading speed texts IReST. *Investigative Ophthalmology & Visual Science*, 53(9):5452–5461, 2012.

[279] S. Trepte and P. K. Masur. Cultural differences in social media use, privacy, and self-disclosure: Research report on a multicultural study. 2016. https://opus.uni-hohenheim.de/volltexte/2016/1218/pdf/Trepte_Masur_ResearchReport.pdf.

[280] S. Trepte, L. Reinecke, N. B. Ellison, O. Quiring, M. Z. Yao, and M. Ziegele. A cross-cultural perspective on the privacy calculus. *Social Media+Society*, 3(1). https://doi.org/10.1177/2056305116688 035, 2017.

[281] H. C. Triandis, R. Bontempo, M. J. Villareal, M. Asai, and N. Lucca. Individualism and collectivism: Cross-cultural perspectives on self-ingroup relationships. *Journal of Personality and Social Psychology*, 54(2):323–338, 1988.

[282] F. Trompenaars and C. Hampden-Turner. *Riding the Waves of Culture: Understanding Diversity in Global Business*. Nicholas Brealey International, 2011.

[283] E. D. Tunstall. *Decolonizing Design: A Cultural Justice Guidebook*. MIT Press, 2023.

[284] J. J. Van Berkum, B. Holleman, M. Nieuwland, M. Otten, and J. Murre. Right or wrong? The brain's fast response to morally objectionable statements. *Psychological Science*, 20(9):1092–1099, 2009.

[285] R. Velt, S. Benford, and S. Reeves. Translations and boundaries in the gap between HCI theory and design practice. *ACM Transactions on Computer-Human Interaction (TOCHI)*, 27(4):1–28, 2020.

[286] T. Victor, K. Kusano, T. Gode, R. Chen, and M. Schwall. Safety performance of the Waymo rider-only automated driving system at one million miles, 2023. https:// storage.googleapis.com/ waymo-uploads /files/documents /safety /Safety%20 Performance%20of%20Waymo%20RO%20at%201M%20miles.pdf.

[287] S. Vieweg and A. Hodges. Surveillance & modesty on social media: How Qataris navigate modernity and maintain tradition. In *Proceedings of the 19th ACM Conference on Computer-Supported Cooperative Work & Social Computing*, pages 527–538, 2016.

[288] D. Vrandečić. In focus: Multilingual Wikipedia, 2020. https://en.wikipedia.org/wiki/Wikipedia:Wikipedia_Signpost/2020-04-26/In_focus.

[289] W. W. W. C. (W3C). World content accessibility guidelines, 2023. https://www.w3 .org/TR/WCAG21/.

[290] W4. Zhihu: Adding "salt value" to your B2B business, May 2022. https://blog .marketingblatt.com/en/zhihu-adding-salt-value-for-your-b2b-business.

[291] M. L. Walters, K. Dautenhahn, R. Te Boekhorst, K. L. Koay, C. Kaouri, S. Woods, C. Nehaniv, D. Lee, and I. Werry. The influence of subjects' personality traits on personal spatial zones in a human-robot interaction experiment. In *ROMAN 2005: IEEE International Workshop on Robot and Human Interactive Communication*, pages 347–352. IEEE, 2005.

[292] H.-C. Wang, S. F. Fussell, and L. D. Setlock. Cultural difference and adaptation of communication styles in computer-mediated group brainstorming. In *Proceedings of the SIGCHI Conference on Human Factors in Computing Systems*, pages 669–678, 2009.

[293] H.-C. Wang, S. R. Fussell, and D. Cosley. From diversity to creativity: Stimulating group brainstorming with cultural differences and conversationally-retrieved pictures. In *Proceedings of the ACM 2011 Conference on Computer Supported Cooperative Work*, pages 265–274, 2011.

[294] L. Wang, P.-L. P. Rau, V. Evers, B. K. Robinson, and P. Hinds. When in Rome: The role of culture & context in adherence to robot recommendations. In *2010 5th ACM/IEEE International Conference on Human-Robot Interaction (HRI)*, pages 359–366. IEEE, 2010.

[295] X. Wang and B. Gu. The communication design of WeChat: Ideological as well as technical aspects of social media. *Communication Design Quarterly Review*, 4(1):23–35, 2016.

[296] Y. Wang, Y. Li, X. Gui, Y. Kou, and F. Liu. Culturally-embedded visual literacy: A study of impression management via emoticon, emoji, sticker, and meme on social media in China. *Proceedings of the ACM on Human-Computer Interaction*, 3(CSCW):1–24, 2019.

[297] Y. Wang, G. Norice, and L. F. Cranor. Who is concerned about what? A study of American, Chinese and Indian users' privacy concerns on social network sites (short paper). In *Proceedings, Trust and Trustworthy Computing: 4th International Conference*, pages 146–153. Springer, 2011.

[298] C. Ward, S. Bochner, and A. Furnham. *The Psychology of Culture Shock*. Routledge, 2020.

[299] C. Ward, Y. Okura, A. Kennedy, and T. Kojima. The U-curve on trial: A longitudinal study of psychological and sociocultural adjustment during cross-cultural transition. *International Journal of Intercultural Relations*, 22(3):277–291, 1998.

[300] C. A. Ward, S. Bochner, and A. Furnham. *The Psychology of Culture Shock*. Routledge, 2001.

[301] B. Warf and P. Vincent. Multiple geographies of the Arab Internet. *Area*, 39(1):83–96, 2007.

[302] S. Warsi. Polygamists are using an app to find their "second wife," September 2016. https://www.vice.com/en/article/nz7qbg/polygamy-secondwife-app-tech.

[303] O. M. Watson and T. D. Graves. Quantitative research in proxemic behavior 1. *American Anthropologist*, 68(4):971–985, 1966.

[304] WeDoJapan. Letters, emails, and the Japanese weather, May 2018. https://www .wedojapan.com/letters-emails-and-the-japanese-weather/.

[305] K. Wenzel, N. Devireddy, C. Davison, and G. Kaufman. Can voice assistants be microaggressors? Cross-race psychological responses to failures of automatic speech recognition. In *Proceedings of the 2023 CHI Conference on Human Factors in Computing Systems*, pages 1–14, 2023.

[306] Wikipedia. Wikipedia:wikipedians. https://en.wikipedia.org/wiki/Wikipedia:Wiki pedians.

[307] Wikipedia. jd.com. https://en.wikipedia.org/wiki/JD.com.

[308] Wikipedia. Wikipedia:Talk page guidelines. https://en.wikipedia.org/wiki/Wikipedia: Talk_page_guidelines.

[309] J. Winawer, N. Witthoft, M. C. Frank, L. Wu, A. R. Wade, and L. Boroditsky. Russian blues reveal effects of language on color discrimination. *Proceedings of the National Academy of Sciences*, 104(19):7780–7785, 2007.

[310] L. Winner. Do artifacts have politics? *Daedalus*, 109(1):121–136, 1980.

[311] L. Winner. *The Whale and the Reactor: A Search for Limits in an Age of High Technology*. University of Chicago Press, 2010.

[312] H. Winschiers-Theophilus and N. J. Bidwell. Toward an Afro-centric indigenous HCI paradigm. *International Journal of Human-Computer Interaction*, 29(4):243–255, 2013.

[313] H. A. Witkin, C. A. Moore, D. R. Goodenough, and P. W. Cox. Field-dependent and field-independent cognitive styles and their educational implications. *Review of Educational Research*, 47(1):1–64, 1977.

[314] Worldbank. Individuals using the Internet (% of population)—Bangladesh, 2021. https://data.worldbank.org/indicator/IT.NET.USER.ZS?locations=BD.

[315] Y. Wu. Is automated journalistic writing less biased? An experimental test of auto-written and human-written news stories. *Journalism Practice*, 14(8):1008–1028, 2020.

[316] W. W. Xu, J. Y. Park, J. Y. Kim, and H. W. Park. Networked cultural diffusion and creation on YouTube: An analysis of YouTube memes. *Journal of Broadcasting & Electronic Media*, 60(1):104–122, 2016.

[317] W. W. Xu, J.-Y. Park, and H. W. Park. Longitudinal dynamics of the cultural diffusion of Kpop on YouTube. *Quality & Quantity*, 51:1859–1875, 2017.

[318] J. Yaaqoubi and K. Reinecke. The use and usefulness of cultural dimensions in product development. In *Extended Abstracts of the 2018 CHI Conference on Human Factors in Computing Systems*, pages 1–9, 2018.

[319] J. Yaaqoubi, Y. Zhang, S. Munson, and K. Reinecke. Barriers in industry that prevent cultural adaptation of web-based products, 2024. In preparation.

[320] J. Yang, M. Morris, J. Teevan, L. Adamic, and M. Ackerman. Culture matters: A survey study of social Q&A behavior. In *Proceedings of the International AAAI Conference on Web and Social Media*, volume 5, pages 409–416, 2011.

[321] T. Yeh, T.-H. Chang, and R. C. Miller. Sikuli: Using GUI screenshots for search and automation. In *Proceedings of the 22nd Annual ACM Symposium on User Interface Software and Technology*, pages 183–192, 2009.

[322] E. Yoon, C.-T. Chang, S. Kim, A. Clawson, S. E. Cleary, M. Hansen, J. P. Bruner, T. K. Chan, and A. M. Gomes. A meta-analysis of acculturation/enculturation and mental health. *Journal of Counseling Psychology*, 60(1):15–30, 2013.

[323] I. M. Young. *Justice and the Politics of Difference*. Princeton University Press, 1990.

[324] Z. Zhang. A guide to using WeChat emojis, July 2016. https://zarazhang.com/2016/07/27/a-guide-to-using-wechat-emojis/.

[325] R. Zhou, J. Hentschel, and N. Kumar. Goodbye text, hello emoji: Mobile communication on WeChat in China. In *Proceedings of the 2017 CHI Conference on Human Factors in Computing Systems*, pages 748–759, 2017.

INDEX

Page numbers in *italic* type refer to figures.

Abokhodair, Norah, 96–97
absolute frame of reference, 55–59
A/B tests, 119, 186
acculturative stress: adjustment stage and, 147; cultural diffusion and, 148–51; culture shock and, *146*, 147–51, 157–59
adjustment stage, 147
affective grounding, 128–33
African Americans, 195–96
Agarwal, Utsav, 161–62
agency: ancient Greeks and, 49; choice and, 67–74; decision-making and, 67–74, 87
Ahmed, Ishtiaque, 165–66
AIM, 124
≠Akhoe Hai//om, 58
algorithms, 111–12, 162, 180, 186
Algorithms of Oppression: How Search Engines Reinforce Racism (Noble), 180
Ali, Syed Mustafa, 188
AliPay, 178–79
allocentric behavior, 66
Alter, Adam, 16
Amazon, 33, 37–42, 75
Anderson, Ken, 93
anthropologists: country borders and, 19; cultural cycles and, 17; culture shock and, 145–46; decision-making and, 62, 66; Gelfand, 138; Hall, 18, 62–63, 121–22, 134, 136; Hofstede, 18–20, 31n1, 41, 77, 81, 105–8; honeymoon stage and, 146; individualism and, 18; Inglehart, 18, 154, *156*; norms and, 121, 138; Oberg, 146; othering and, 15, 188; Pelto, 18, 138–39, 141; as research source, 11–12, 15; Schwartz, 18, 68; Triandis, 18,

66, 89; Trompenaar, 18; values and, 15, 17–20, 138, 145
anxiety stage, 146–47
Apple, 125, 127, 183
Arnett, Jeffrey, 8
artificial intelligence (AI): bias and, 168, 181; chatbots and, 32, 151–57, 204; ChatGPT, 143, 151, 153, 168, 171, 180, 195–200, 204; China and, 21; communication and, 127; context and, 41; culture shock and, 151–53, 157–60; decision-making and, 201; democracy and, 181; digital juries, 176–77; diversity and, 198–200; education and, 180–81; ethics and, 9–10; Genmoji and, 127; Google and, 167–68; homogenization and, 200; humanization of, 41; hype of, 2; India and, 168–69; infrastructure and, 180–82, 195–97; Japan and, 2; language and, 167–71, 196; large language models and, 151, 158, 167, 170–71; marginalization and, 167–71; moral issues and, 9, 154; news publications and, 181; norms and, 127; Personality Design Team and, 41; religion and, 168; robots and, 2, 9, 151, 153, 158, 195; SenseTime and, 21; sentience and, 153; social justice and, 182; societal threat of, 157–60; trust of, 2–3; values and, 153, 195, 198, 200, 204
attention patterns: decision-making and, 64; distractions and, 2, 24, 51, 90; information and, 45–52, 58–60, 201; norms and, 137
Australia, 20; Canva and, 21; cultural background and, 117; decision-making and, 66; Guugu Yimithirr and, 57; norms

Australia (*continued*)
and, 145; online communities and,
91; Pormpuraaw and, 55–57; robots and,
2; website preferences of, 117
Austria, 63, 116, 175
autonomous vehicles: context and, 201;
robotaxis, 1–10, 35; Sensetime and, 21
autonomy, 67–68, 76, 99, 154
Awad, Edmond, 9
Axim Collaborative, 84

Badre, A., 105
Baidu, 27, 178
Balinese, 57
Bangladesh: khep and, 164–65; marginalization and, 161–68, 171–74; religion
and, 165–66; Uber and, 161–67
banks, 36, 48, 103, 178–79
Barber, W., 105
Basoah, Jeffrey, 195
behavior: allocentric, 66; average, 17;
choice and, 4, 67, 78, 131, 141; cognition
and, 22, 136; communication and, 122–
23, 131–32; complexity of human, 8;
context and, 27–31, 201; cultural cycles
and, 17; culture shock and, 145, 151,
153, 159; decision-making and, 66–67,
71, 78–82; driving, 3–4; external validity
and, 8; idiocentric, 66; indulgence vs.
restraint, 31n1; information and, 45–46,
51; infrastructure and, 194–95; norms
and, 11, 15–17, 29, 71, 122–23, 131–38,
141–42, 194, 201; psychologists and, 8
(*see also* psychologists); robots and, 3;
technology creators and, 7; values and,
15, 17, 22, 71, 122, 153, 159, 194–95,
201
Behavioral and Brain Sciences journal, 8–9
Belhare, 57
Belize, 58
Benartzi, Shlomo, 76–77
Bengali, 162–63, 171–76
Bennett, Cynthia, 168
Bernstein, Abraham (Avi), 106
Berry, John Widdup, 147
Beyer, Hugh, 34
bias: artificial intelligence (AI) and, 168,
181; communication and, 127; context
and, 33; culture shock and, 151–52,

155–58; decision-making and, 83;
designers and, 117–19; implicit, 20;
information and, 59; infrastructure and,
181, 188–89; marginalization and, 168–
77; news, 181; norms and, 20–21, 127;
participant samples and, 8–12, 152; tech
hubs and, 20–21, 197; website design
and, 117–19; Western, 11, 20, 33, 59, 83,
168, 181, 188–89
Bidwell, Nicola, 99–100, 189
bike-sharing, 179
BlenderBot, 153–55
Bochner, Stephen, 145
Boroditsky, Lera, 55–57
Bosnia-Herzegovina, 114, 117
Brazil: decision-making and, 63; infrastructure and, 186; JingDong and, 21;
marginalization and, 169; online communities and, 94; Portuguese and, 41;
Tiriyó and, 58; Wildlife Studio and, 21
Britton, Lauren, 95
Buddhism, 173
ByteDance, 21

Canada: cultural background and, 112,
117; culture shock and, 147; decision-
making and, 66; French, 41; gestures
and, 122; norms and, 122, 130–31;
website preferences of, 117
Canva, 21
capitalism, 7, 14, 29
CashApp, 36
Catholic Church, 9n1, *156*
CB Insights, 20
censorship, 153, 166, 174
Centre for Social Change, 165
chatbots: culture shock and, 32, 151–57,
204; sentience and, 153; values and, 153
ChatGPT: artificial intelligence (AI) and,
143, 151, 153, 168, 171, 180, 195–200,
204; communication and, 143; culture shock and, 151, 153; diversity and,
198–200; education and, 180–81; infrastructure and, 180–81, 195–96; large
language models and, 151, 158, 167,
170–71; marginalization and, 168, 171,
200; misassumptions on, 204; norms
and, 143; values and, 153, 195, 198, 200,
204

Chau, Patrick, 89
CHI journal, 11
Chile, 21, 63, 114
China: artificial intelligence (AI) and, 21; Baidu and, 27, 178; ByteDance and, 21; communication and, 125–29, 133–38, 141–42; Confucianism and, 49; context and, 27, 31, 40; cultural background and, 105, 108, 119; decision-making and, 65, 66, 68, 73–77, 84; Douyin and, 21; e-commerce and, 21; entrepreneurs and, 22; gestures and, 122; hegemony and, 200; iCourse and, 84; information and, 47–50, 52; infrastructure and, 178–79, 197; innovation and, 21–22; JingDong and, 21; Mandarin and, 124, 178; marginalization and, 165–66; norms and, 122, 125, 142; online communities and, 89–92; politics and, 21–22; SenseTime and, 21; Taobao and, 75; tech hubs of, 21; TikTok and, 21; WeChat and, 21, 125–26, 178–79; Weibo and, 89; XuetangX and, 84; Yu Garden, 178; Zhihu and, 89, 91–92
Chiumbu, Sarah, 181
choice: agency and, 67–74; autonomy and, 67–68, 76, 154; behavior and, 4, 67, 78, 131, 141; capitalistic, 14; conformity and, 69, 142; context and, 30, 32, 34, 43; cultural background and, 67–74, 102–8, 113, 117–19; culture shock and, 154; decision-making and, 67–84, 87, 202; design, 12, 34, 43, 76, 108, 113, 118–19; freedom of, 67–68, 72, 77, 79; infrastructure and, 194; interdependency and, 71–76, 87, 201; Massive Open Online Courses (MOOCs) and, 78–86, 201–2; misassumptions on, 202; moral issues and, 9, 67, 194; norms and, 125–28, 131, 133, 141, 202; online communities and, 94; rewards and, 69; search engines and, 30; self-inflation and, 72–73; web navigation and, 75–84
choice overload: being overwhelmed by, 68, 75; cultural differences and, 75–79; navigating the web and, 75–84; recommendations and, 75–76
Christianity, 153–54, 173, 175

classic alienation, 147
co-designers, 190
code-switching, 137–44, 194
cognition: analytic vs. holistic, 46–47; attention patterns and, 45–52, 58–60; behavior and, 22, 136; brain plasticity and, 12; cultural background and, 45–54; decision-making and, 81; distractions and, 2, 24, 51, 90; eye movements and, 47–49; foreground effects and, 45–51; hippocampus and, 44–45; information and, 12, 45–46, 55, 58–60, 143; interaction paradigms and, 59–60; Kitayama and, 51–52, 71–72, 74; Maguire study and, 44–45; neural costs of, 202; norms and, 136, 143; online communities and, 100; priming and, 50; reading speed and, 48
collectivism: communication and, 132; conformity and, 69, 142; context and, 25, 31–32, 36, 41; cultural background and, 105–6; culture shock and, 157–60; decision-making and, 64–66, 76, 202; democracy and, 18; Doodle and, 64–66; Hofstede and, 18–19; individualism and, 18, 31, 41, 66, 76, 91, 94, 97–99, 105–6, 202; norms and, 132; online communities and, 89–99; social Darwinism and, 9n1; Tocqueville and, 18
colonialism: border issues and, 15; British, 99; imperialism and, 22; infrastructure and, 187–90; marginalization and, 164–65, 170–71, 174–76; neocolonialism, 13–14, 100, 165; online communities and, 99–100; postcolonialism, 13–14, 188
color: context and, 25, 29, 33, 43; cultural background and, 25, 29, 33, 43, 102–7, 111–19; decision-making and, 69, 76; designers and, 12, 29, 43, 102–7, 111–19, 171, 184–87, 198; information and, 58; infrastructure and, 184–87; language and, 58; online communities and, 92; user interfaces and, 12, 25, 33, 76, 102–3, 105, 111, 184, 186; websites and, 25, 33, 103, 105, 107, 111–19, 186–87
communication: across cultures, 123–33; affective grounding and, 128–33; artificial intelligence (AI) and, 127;

communication (*continued*)
  behavior and, 122–23, 131–32;
    bias and, 127; contact cultures and
    135, 143; censorship of, 153, 166,
    174; chatbots and, 32, 151–57, 204;
    ChatGPT and, 143; China and, 125–29,
    133–38, 141–42; code-switching, 137–
    44; collectivism and, 132; context and,
    121–25, 129, 132–35, 137, 139, 142–43,
    203; designers and, 126; diversity and,
    132, 142–43; efficiency and, 120–22,
    129; email, 120–21, 130–31; emojis,
    121, 125–29; emoticons, 124–25, 129;
    equality and, 122; facial expressions,
    122–23, 195, 203; feedback, 130–33,
    190; formal, 41, 120–23, 131, 185, 203;
    gestures, 122–23, 126, 134, 136, 143,
    145, 195, 203; Google and, 127; high-
    context cultures and, 122-24, 129, 134,
    143, 203; homogenization and, 127,
    142, 198; Human-Computer Interaction
    (HCI), 126, 130; identity and, 125, 128;
    instant messengers, 94, 120, 123–25,
    128, 130; intergroup threat and, 135–37,
    195; Japan and, 120, 123–27, 130–32,
    135–37; LabintheWild and, 121, 138–
    39; language and, 121–24, 127, 131–32,
    135, 138; loose cultures and, 138–41;
    low-context cultures and, 122, 125, 132–
    33, 137, 143, 203; noncontact cultures
    and, 135, 143; politics and, 138–39;
    power distance and, 122–25; proxemics
    and, 134–35; psychologists and, 124,
    128, 130, 135, 138; reading between
    lines, 121; religion and, 139; robots and,
    133–44, 203–4; silence and, 122–23,
    143, 203; Slack and, 120–21, 123–24,
    131, 143, 148; streamlining, 120; tight
    cultures and, 138–42; values and, 122,
    138
conformity, 69, 142
Confucianism, 49
contact cultures: communication and,
    135, 143; culture shock and, 145–46;
    misassumptions on, 203
context: analytic vs. holistic, 46–47; artifi-
    cial intelligence (AI) and, 41; behavior
    and, 27–31, 201; bias and, 33; China
    and, 27, 31, 40; choice and, 30, 32, 34,
    43; cluttered cities and, 50–51; collec-
    tivism and, 25, 31–32, 36, 41; color and,
    25, 29, 33, 43; communication and,
    121–25, 129, 132–35, 137, 139, 142–
    43, 203; corporate culture, 20; cultural
    dimensions and, 41; culture shock and,
    149, 158; decision-making and, 85, 87;
    designers and, 27, 34–36, 39; develop-
    ers and, 27, 29, 36–37; diversity and,
    28–31; Doodle and, 17; e-commerce
    and, 32–40; efficiency and, 24, 27, 39;
    external validity and, 8; Facebook and,
    30, 32; foreground and, 45–51; gen-
    der and, 30, 32, 41; Germany and, 31,
    36, 39, 62; Google and, 24–29, 37, 40;
    Hall and, 121–22; Human-Computer
    Interaction (HCI) and, 28; identity and,
    30; India and, 28, 39–41; individualism
    and, 30–31, 41; indulgence vs. restraint,
    31n1; information and, 45–53, 57, 59;
    infrastructure and, 185, 189, 192–93;
    ingroups and, 25, 31; innovation and,
    35; Japan and, 27–31, 42; labels and, 16;
    LabintheWild and, 31; language and, 28,
    37–41; limited research samples, 8; local,
    6, 27, 30, 36–43, 85, 87, 172, 176, 189;
    marginalization and, 169, 172, 176–77;
    Microsoft and, 40–42; misassumptions
    on, 201; neuroscience and, 45; norms
    and, 121–25, 129, 132–39, 142–43,
    203; online communities and, 29–34,
    93–97, 101; participant samples and,
    9–11; political, 21, 27, 30; priming and,
    50; religion and, 36; rewards and, 33;
    robotaxis and, 4, 6; search engines and,
    24–30, 37, 41; self-driving cars and, 201;
    social media and, 30–31; South Korea
    and, 24–27, 37; specific cultural, 35–37;
    Sweden and, 42; Switzerland and, 39;
    transplanting, 6, 201; United Kingdom
    and, 31, 36; United States and, 24, 27,
    29–34, 37–41; user interfaces and, 24–
    25, 27, 33; websites and, 25, 27, 33–34,
    37–42
Contextual Design (Beyer and Holtzblatt),
    34
corporate culture, 20
Cortana, 40–42
Coupang, 21

Coursera, 78, 84
COVID-19 pandemic, 179
credit cards, 161
crowdsourcing, 152
culturability, 105
cultural appropriation, 93, 137, 182, 196
cultural background: Canada and, 112, 117;
    China and, 105, 108, 119; choice and,
    67–74, 102–8, 113, 117–19; cognition
    and, 45–54; collectivism and, 105–6;
    color and, 25, 29, 33, 43, 102–7, 111–
    19; cultural dimensions and, 105–8;
    designers and, 108, 110, 119, 189–90;
    education and, 113–14; efficiency and,
    39, 102, 110; gender and, 105, 113–14,
    118; gestures and, 122, 126; Google and,
    117; information and, 45–54; infras-
    tructure and, 189–90; interdependency
    and, 105; Japan and, 105; LabintheWild
    and, 109–13, 200; learning styles and,
    33, 58, 77, 87; localization and, 107;
    monochronic vs. polychronic, 61; power
    distance and, 105, 107; quantifying
    first impressions and, 111–17; rituals
    and, 41, 96, 166, 198; search engines
    and, 117; Sweden and, 116; Switzerland
    and, 102–8, 116; Uber and, 162–63;
    uncertainty avoidance and, 106; United
    Kingdom and, 117; United States and,
    107–10, 117–18; user interfaces and,
    102–11, 118; values and, 106; websites
    and, 103–19
cultural cycles: anthropologists and, 17;
    information and, 12; national culture
    and, 20; psychologists and, 17; technol-
    ogy and, 17–18; tight cultures and, 154;
    values and, 12, 17–20, 194
cultural diffusion, 148–51
cultural dimensions: context and, 41;
    cultural background and, 105–8;
    decision-making and, 77; Hofstede
    and, 18–20, 41, 77, 105–8; labels and,
    18; Tocqueville and, 19
cultural exchanges, 145
culturally intelligence, 142–44
cultural markers, 105, 168
culture shock: acculturative stress and,
    146, 147–51, 157–59; adjustment stage
    and, 147; anthropologists and, 145–46;

anxiety stage and, 146–47; artificial
    intelligence (AI) and, 151–53, 157–60;
    behavior and, 145, 151, 153, 159; bias
    and, 151–52, 155–58; Canada and, 147;
    chatbots and, 32, 151–57, 204; Chat-
    GPT and, 151, 153; choice and, 154;
    classic alienation and, 147; collectivism
    and, 157–60; contact cultures and,
    145–46; context and, 149, 158; cultural
    diffusion and, 148–51; digital, 5–6, 13,
    148, 201, 204; efficiency and, 151; equal-
    ity and, 148, 152, 155; Facebook and,
    149–50; gender and, 148, 152, 155; ges-
    tures and, 145; Google and, 153; health
    and, 147, 157–58; homogenization and,
    159; honeymoon stage and, 146; ideals
    and, 157; identity and, 147, 150, 152;
    India and, 149; individualism and, 157–
    60; ingroups and, 149; LabintheWild
    and, 152–53; language and, 145, 152–53,
    158; marginalization and, 5–6; mis-
    assumptions on, 201–4; politics and,
    150–54; power shock and, 149; psychol-
    ogists and, 145–47, 152; rejection stage
    and, 147; religion and, 148, 152–55;
    rewards and, 145; robots and, 151, 153,
    158; safety and, 159; self-esteem and,
    147, 158, 200; self-expression and, 154–
    55; similarity-attraction theory and, 150;
    social media and, 148–49, 157; societal
    threat of, 157–60; survival and, 154–55;
    Sweden and, 155; United Kingdom and,
    155; United States and, 149–50, 153,
    155; values and, 13, 146–60
Czech Republic, 145

DALL-E, 167
Das, Dipto, 171–73, 176
decision-making: agency and, 67–74, 87;
    anthropologists and, 62, 66; artificial
    intelligence (AI) and, 201; attention and,
    64; autonomy and, 67–68, 76; behavior
    and, 66–67, 71, 78–82; bias and, 83;
    Brazil and, 63; Canada and, 66; China
    and, 65, 66, 68, 73–77, 84; choice and,
    66–84, 87, 202; cognition and, 81; col-
    lectivism and, 64–66, 76, 202; color and,
    69, 76; conformity and, 69, 142; context
    and, 85, 87; cultural

decision-making (*continued*)
　background and, 67–74; cultural dimen-
　sions and, 77; developers and, 84; diver-
　sity and, 76, 80; Doodle and, 62–67, 73;
　education and, 68, 70, 77–86; efficiency
　and, 62–63; equality and, 87; Germany
　and, 62–63; health and, 78; homogeniza-
　tion and, 67; Hong Kong and, 66–68;
　India and, 63–67; individualism and, 66,
　76; ingroups and, 66, 70–71, 75–76, 202;
　interdependency and, 71–76, 87, 201;
　Japan and, 66, 71–73; LabintheWild and,
　66, 72; learning styles and, 78–84; Mas-
　sive Open Online Courses (MOOCs)
　and, 78–86, 201–2; misassumptions on,
　202; moral issues and, 67–68; overload
　and, 75–84; power shock and, 77, 83;
　psychologists and, 62, 64, 66–71, 81;
　rewards and, 69, 76; search engines and,
　76; self-inflation and, 72; social media
　and, 76; Sweden and, 77, 84; Switzerland
　and, 62–63; time and, 62–67; uncer-
　tainty avoidance and, 77–78; United
　Kingdom and, 72–73; United States and,
　62–84; user interfaces and, 76; values
　and, 81, 85; web navigation and, 75–84;
　websites and, 78
Declaration of Independence, 67–68
decolonization, 187–90
Deliveroo, 21
democracy: artificial intelligence (AI) and,
　181; collectivism and, 18; education and,
　79, 83; equality and, 18; misassumptions
　on, 204; tech giants and, 180; values and,
　148, 152–53, 155, 204
Democracy in America (de Tocqueville), 18
Denmark, 21, 77, 116
Denton, Emily, 168
designers: bias and, 117–19; color and,
　12, 29, 43, 102–7, 111–19, 171, 184–
　87, 198; communication and, 126;
　context and, 27, 34–36, 39; cultural
　background and, 108, 110, 119; cul-
　tural diversity and, 11–23; decolonizing,
　187–90; Google and, 24–29, 37, 40,
　117, 183; IDEO and, 34–35; infrastruc-
　ture and, 14, 182, 186–93; localization
　and, 36–43, 193–97; Massive Open
　Online Courses (MOOCs) and, 78;

minimalism and, 27, 114; misassump-
　tions on, 201–2; norms and, 6, 12–13,
　34, 126, 189, 193; online communities
　and, 91, 93; participatory design and, 14,
　190–92; postcolonial computing and,
　14, 188; streamlining and, 39, 102, 120;
　values and, 6, 12–13, 187–88; website, 7,
　34, 39–42, 54, 78, 100–19, 171, 183–87
developers: context and, 27, 29, 36–37;
　decision-making and, 84; infrastructure
　and, 14, 184, 187, 190–96; marginaliza-
　tion and, 166–67, 173, 176; norms and,
　6, 12–13, 29, 133, 143, 166, 193; values
　and, 6, 12–13, 187, 195–96
digital juries, 176–77
distractions, 2, 24, 51, 90
diversity: artificial intelligence (AI) and,
　198–200; ChatGPT and, 198–200;
　communication and, 132, 142–43; con-
　text and, 28–31; creativity and, 204;
　decision-making and, 76, 80; design-
　ing for, 11–23; Human-Computer
　Interaction (HCI) and, 11–12; ideals
　and, 28, 170; information and, 58, 60;
　infrastructure and, 181, 191, 196–97;
　marginalization and, 168, 170, 173; mis-
　assumptions on, 204; norms and, 132,
　142–43; retaining, 198
Dong, Ying, 47
Doodle: collectivism and, 64–66; data logs
　and, 63; decision-making and, 62–67,
　73; Germany and, 17; hidden polls of,
　64; Levine and, 63; Naef and, 62–63;
　time-perception and, 62–67
Dosono, Bryan, 95
Dourish, Paul, 93, 180
Douyin, 21
Drunk Tank Pink (Alter), 16
Dutch, 58

early adopters, 162
e-commerce: China and, 21; complex
　websites and, 109; context and, 32–40;
　mobile payments and, 178–79
Ecuador, 21, 92
Edraak, 78
education: artificial intelligence (AI) and,
　180–81; Axim Collaborative and, 84;
　certificate earners and, 80; ChatGPT

and, 180–81; cultural background and, 113–14; decision-making and, 68, 70, 77–86; democratization of, 79, 83; edX and, 78–79, 84–85; IITBombayX and, 84–85; India and, 80–85; infrastructure and, 180; language and, 80; MOOCs and, 78–86, 201–2; One Laptop Per Child Project and, 86; online communities and, 89; STEM courses, 118; student-teacher ratios and, 81–83

Edwards, Paul, 180

edX, 78–79, 84–85

efficiency: communication and, 120–22, 129; context and, 24, 27, 39; cultural background and, 39, 102, 110; culture shock and, 151; decision-making and, 62–63; designers and, 13, 24, 27 (see also designers); impact of, 13; infrastructure and, 181; marginalization and, 164; misassumptions on, 202; norms and, 120–22, 129; online communities and, 89–90

egocentric frame of reference, 55, 59, 94, 99

Egypt, 4–5, 16, 32, 36, 128

Electrolux, 42

El Salvador, 92

email: communication and, 120–21, 130–31; cultural interaction and, 148; infrastructure and, 180; pre-written, 61

emojis, 121, 125–29

emoticons, 124–25, 129

entrepreneurs, 22, 162

equality: communication and, 122; culture shock and, 148, 152, 155; decision-making and, 87; democracy and, 18; gender, 139, 148, 152, 155; misassumptions on, 204; norms and, 122; values and, 122, 148, 152, 155, 194, 204

Eraqi, Hesham, 5

Estonia, 139

ethics: artificial intelligence (AI) and, 9–10; consensus on, 10; homogenization and, 182, 196; infrastructure and, 182, 196; lifeboat task and, 128; norms and, 128; robots and, 9–10; values and, 196, 200; Western, Educated, Industrialized, Rich, and Democratic (WEIRD) people and, 9

Ewe Guugu, 58

external validity, 8

eye movements, 47–49

Facebook: audience selector of, 100; censorship of, 166; context and, 30, 32; culture shock and, 149–50; Groups, 93–94, 162, 176; infrastructure and, 180; language and, 169; marginalization and, 162, 166–69, 176; Namibia and, 95–96; online communities and, 89, 93–95, 100; religion and, 166; Solidarity Economy and, 93; Uber and, 162, 167, 169; WeChat and, 21; Zuckerberg and, 93–94

facial expressions, 122–23, 195, 203

Fan, Jenny, 176–77

feedback, 130–32, 190

Finland, 2, 21, 114, 116, 187

Forbes 500 companies, 105

formality, 41, 120–23, 131, 185, 203

France, 91, 116, 126, 145

freedom of choice, 67, 72, 77, 79

Friedman, Batya, 188

Friendster, 93

Fuhrman, Orly, 55

Furnham, Adrian, 145

Fussell, Susan, 124

Gaby, Alice, 55–57

Gajos, Krzysztof, 113

Gao, Ge, 128

Geek Heresy (Toyama), 85

Gelfand, Michele, 139, 142

gender: context and, 30, 32, 41; cultural background and, 105, 113–14, 118; culture shock and, 148, 152, 155; equality and, 139, 148, 152, 155; identity and, 30; masculinity, 105–7; norms and, 136, 139

Genmoji, 127

Geography of Thought, The (Nisbett), 49

Geography of Time, The (Levine), 62

Georgia Institute of Technology, 93, 126

Germany, 3, 20; context and, 31, 36, 39, 62; cultural exchanges and, 145; decision-making and, 62–63; Doodle and, 17; language and, 80, 175; online communities and, 91–92; self-inflation and, 72; time perception and, 63; website preferences of, 116 gestures: communication and, 122–23, 126, 134, 136, 143, 145, 195, 203; cultural, 122, 126; culture shock and, 145; interpretation of, 122; Italy and, 122; norms and, 122, 126, 134, 136, 143; trust and, 195

Ghana, 58

Global South, 152, 162, 164, 181

Global Tech Hub Report, 20

Google: artificial intelligence (AI) and, 167–68; communication and, 127; context and, 24–29, 37, 40; cultural background and, 117; culture shock and, 153; designers and, 24–29, 37, 40, 117, 183; Imagen, 167; impact of, 180; information and, 44; infrastructure and, 180, 183; Meet, 183; Naver and, 24–25, 26, 29; norms and, 127; online communities and, 88; Raghavan and, 28; Research, 168; Russell and, 25, 27–28; search market share of, 24; Search Quality, 25; sentience and, 153; South Korea and, 24; Western, Educated, Industrialized, Rich, and Democratic (WEIRD) people and, 29, 88

GPT-3, 153–55

Graham, Mark, 170

Gravett, Willem, 165

Greece, 77, 116, 155

Guatemala, 77

Guo, Philip, 78, 81, 83

Gutchess, Angele, 51–52

Guugu Yimithirr, 57

Guyatt, Nicholas, 9n1

Hai//om, 57

Hall, Edward: communication and, 121–22, 134–36; cultural theories and, 18, 121–22; decision-making and, 62–63; *The Silent Language*, 62

halo effect, 109–10

He, Helen, 130–31

health, 133; accessibility issues and, 183–84; culture shock and, 147, 157–58; decision-making and, 78; infrastructure and, 183; religion and, 166

Hecht, Brent, 170–71

Hedden, Trey, 53–55

hegemony, 127, 159, 166, 197, 200

Heine, Steven, 8–9

Henrich, Joseph, 8–9

Herero tribe, 189

high-context cultures: communication and, 122–24, 129, 134, 143, 203; misassumptions on, 203; online communities and, 95; robots and, 134, 142–43, 203

Hinds, Pamela, 123

Hindus, 172–74

hippocampus, 44–45

Hodges, Adam, 97–98

Hofstede, Geert: collectivism and, 18–19; cultural dimensions and, 18–20, 41, 77, 105–8; education and, 81; indulgence vs. restraint and, 31n1; masculinity and, 105–7

Holtzblatt, Karen, 34

homogenization: artificial intelligence (AI) and, 200; communication and, 127, 142, 198; culture shock and, 159; decision-making and, 67; norms and, 127, 142

honeymoon stage, 146

Hong Kong, 21, 52, 66–68, 89

Höök, Kristina, 193

Housing.com, 21

Human-Computer Interaction (HCI): communication and, 126, 130; context and, 28, 34; designers and, 34; diversity and, 11–12; infrastructure and, 180, 187; norms and, 126, 130; Western, Educated, Industrialized, Rich, and Democratic (WEIRD) people and, 11

Hungary, 139

IBM, 19–20

iCourse, 84

ICQ, 124

ideals: academic, 14; culture shock and, 157; diversity and, 28, 170; marginalization and, 170, 176; moderators and, 176; online communities and, 157; pragmatism and, 190–93; social media and, 157; Western, 101

identity: classic alienation and, 147; collective, 95, 98, 203; communication and, 125, 128; context and, 30; culture shock and, 147, 150, 152; gender, 30;

individualism and, 9n1, 18–19, 30–31, 41, 66, 76, 91, 94, 97–101, 105–7, 202–3; ingroups and, 18, 25, 31, 66, 70–71, 75–76, 144, 149, 195, 201–3; intergroup threat and, 135–37, 195; labels and, 16–19, 23, 168; marginalization and, 172–76; norms and, 125, 128; online communities and, 95–101; personal growth and, 200; social media and, 18, 31, 76, 97, 99, 202–3; stereotypes, 13, 16, 62, 122, 130; Twenty-Statements Test and, 98–99

IDEO, 34–35

idiocentric behavior, 66

ideologies, 150, 166, 170

Imagen, 167

imperialism: colonialism and, 22; digital, 165; goal of, 164; marginalization and, 13–14, 165, 167, 169

India: artificial intelligence (AI) and, 168–69; certificate earners and, 80; context and, 28, 39–41; culture shock and, 149; decision-making and, 63–67; education and, 80–85; Housing.com and, 21; language and, 28, 39, 41, 80, 124, 132, 135, 169–72; long search queries of, 28; norms and, 124, 128, 135–36, 139; online communities and, 89–91, 94; Orkut and, 89; Quora and, 89, 171–74; Snapdeal and, 21; Swayam and, 78

individualism: Catholic Church and, 9n1; collectivism and, 18, 31, 41, 66, 76, 91, 94, 97–99, 105–6, 202; context and, 30–31, 41; cultural background and, 105–7; culture shock and, 157–60; decision-making and, 66, 76; Hofstede score and, 19; identity and, 9n1, 18–19, 30–31, 41, 66, 76, 91, 94, 97–101, 105–7, 202; misassumptions on, 202–3; online communities and, 91, 94, 97–101, 202; social media and, 18, 31, 76, 97, 99, 202–3; Tocqueville on, 18; United States and, 19–20

Indonesia, 57

indulgence vs. restraint, 31n1

information: analytic vs. holistic, 46–47; attention patterns and, 45–52, 58–60, 201; behavior and, 45–46, 51; bias and, 59; brain plasticity and, 12; China and, 47–50, 52; cognition and, 12, 45–46, 55, 58–60, 143; color and, 58; context and, 45–53, 57, 59; crosscultural perception of, 44–60; cultural background and, 45–54; cultural cycles and, 12; diversity and, 58, 60; eye movements and, 47–49; foreground and, 45–51; Google and, 44; hippocampus and, 44–45; Hong Kong and, 52; interaction paradigms and, 59–60; interdependency and, 49; Japan and, 46, 48–55; Kitayama and, 51–52, 71–72, 74; LabintheWild and, 52–53; language and, 47, 54–59; London taxi routes and, 44–45; misassumptions on, 201–2; neuroscience and, 44–45, 51; Nisbett study and, 46–50; organizing, 54–58; priming and, 50; psychologists and, 45–46, 49–53, 66–71; reading speed and, 48; South Korea and, 47–48; taxonomic terms and, 27, 49, 59, 201; United Kingdom and, 45; United States and, 46–53, 56–57; user interfaces and, 54–55, 58–60; values and, 51; websites and, 47–48–50, 54

infrastructure: adaptability and, 193–97; artificial intelligence (AI) and, 180–82, 195–97; behavior and, 194–95; bias and, 181, 188–89; Brazil and, 186; building culturally just, 178–97; ChatGPT and, 180–81, 195–96; China and, 178–79, 197; choice and, 194; color and, 184–87; context and, 185, 189, 192–93; cultural background and, 189–90; decolonizing, 187–90; designers and, 14, 182, 186–93; developers and, 14, 184, 187, 190–96; diversity and, 181, 191, 196–97; education and, 180; efficiency and, 181; email and, 180; ethics and, 182, 196; Facebook and, 180; Google and, 180, 183; health and, 183; Human-Computer Interaction (HCI) and, 180, 187; ideals and, 190–93; ingroups and, 195; language and, 182–84, 188–91, 195–97; localization and, 182–97; Microsoft and, 183; mobile payments and, 178–79; moral issues and, 185, 192, 194, 196; mundane technologies and, 180; participatory design and, 14, 190–92; platform dominance of, 180; politics and, 181, 188; postcolonial computing and, 14, 188;

infrastructure (*continued*)
pragmatism and, 190–93; psychologists and, 195; robots and, 188–89, 195; search engines and, 180; social justice and, 182–87; social media and, 182; taxis and, 179; United Kingdom and, 187–88; United States and, 178, 181–83, 195–97; user interfaces and, 184, 186, 191, 193–95; WCAG and, 183–85; websites and, 183–87
Inglehart, R., 18, 154, *156*
ingroup condition, 70–71
ingroups: context and, 25, 31; culture shock and, 149; decision-making and, 66, 70–71, 75–76, 202; identity and, 18, 25, 31, 66, 70–71, 75–76, 144, 149, 195, 201–3; infrastructure and, 195; norms and, 144; outgroups and, 25, 70–71, 76
innovation: China and, 21–22; context and, 35; disruptions from, 161–62; online communities and, 93; Uber and, 161–62
instant messengers, 94, 120, 123–25, 128, 130
interdependency: choice and, 71–76, 87; cultural background and, 105; decision-making and, 71–76, 87, 201; information and, 49; online communities and, 95, 203
intergroup threat, 135–37, 195
Irani, Lilly, 188
Iraq, 95
Ireland, 63
isiZulu, 197
Italy, 39, 116, 122
Iyengar, Sheena, 70

Jafarinaimi, Nassim, 162
Jamaica, 77
Japan, 22; analytic vs. holistic cognition and, 46–47; artificial intelligence (AI) and, 2; cluttered cities and, 50–51; communication and, 120, 123–27, 130–32, 135–37; context and, 27–31, 42; cultural background and, 105; decision-making and, 66, 71–73; emojis and, 125–27; Forbes 500 companies and, 105; information and, 46, 48–55; Mercari and, 21; Mixi and, 30–32; norms and, 120,

123–27, 130–32, 135–37; online communities and, 94–95; reading speed and, 48; SoftBank and, 27; succinct search queries of, 28; Yahoo! and, 26, 27, 29
JingDong, 21
Jordan, 78
Journal of Consumer Research, 68
Journal of Personality and Social Psychology, 70
Joy of Search, The (Russell), 28
Jumia, 32–33
Jung, Malte, 133
Justice and the Politics of Difference (Young), 165

KAIST, 47
Kenya, 81–83, 99
khep, 164–65
Kilivila, 58
Kim, Heejung, 68–69
Kimura-Thollander, Philippe, 126–27
Kitayama, Shinobu, 51–52, 71–72, 74
"Knowledge, The" (London taxi exam), 44–45
Komplett.no, 33
Konga, 32–33
Kosinski, Michal, 94
Kuhn, Manford, 98–99
Kumar, Neha, 126–27, 162
Kuuk Thaayorre, 55–57
Kwet, Michael, 165

labels, 16–19, 23, 168
LabintheWild: communication and, 121, 138–39; context and, 31; cultural background and, 109–13, 200; culture shock and, 152–53; decision-making and, 66, 72; information and, 52–53; norms and, 121, 138–39; online communities and, 98; purpose of, 22–23; Western, Educated, Industrialized, Rich, and Democratic (WEIRD) people and, 23, 200
Lagoze, Carl, 180
language: artificial intelligence (AI) and, 167–71, 196; as barrier, 91, 135, 145, 163, 169–77, 184; ChatGPT and, 151; color and, 58; communication and,

121–24, 127, 131–32, 135, 138; context and, 28, 37–41; Cortana and, 41; culture shock and, 145, 152–53, 158; education and, 80; Facebook and, 169; Germany and, 80, 175; Hall on, 62, 121–22; India and, 28, 39, 41, 80, 124, 132, 135, 169–72; information and, 47, 54–59; infrastructure and 182–84, 188–91, 195–97; interaction paradigms and, 59–60; large language models and, 151, 158, 167, 170–71; linguistic relativity and, 58; localization and, 42–43; marginalization and, 162–63, 169–77; moderators and, 172–77; natural language processing and, 152; norms and, 121–22, 124, 127, 131–32, 135, 138; Quora and, 172–76; reference frames and, 55–58; *The Silent Language*, 62; spatial references and, 55–57; subcultural groups and, 15; Switzerland and, 175; translation of, 42; Uber and, 162–63, 169; Wikipedia and, 169–71

large language models, 151, 158, 167, 170–71

Law of Amplification, 86

learning styles: cooperation, 33; cultural background and, 33, 58, 77, 87; decision-making and, 78–84; field-dependent learners, 81; field-independent learners, 81–82; Maguire study, 45; MOOCs, 86 (*see also* Massive Open Online Courses (MOOCs)); self-directed, 77; spatial tasks, 58

Lebanon, 17, 63

Lee, Kun-Pyo, 47

Lepper, Mark, 70

Levine, Robert V., 62–63

lifeboat task, 128

Lindgaard, Gitte, 112

Lindtner, Silvia, 22, 93

linguistic relativity, 58

Liu, Shuqing, 84

localization: context and, 27, 36–43; Cortana and, 40–42; cultural background and, 107; design, 36–43, 193–97; infrastructure and, 182–97; language and, 42–43; marginalization and, 167, 169; user, 193

Longgu, 57

loose cultures, 95, 138–41

low-context cultures: communication and, 122, 125, 132–33, 137, 143, 203; English Talk and, 132; explicit nature of, 122; Hall and, 122; misassumptions on, 203; online communities and, 125; robots and, 133–34, 137, 143, 203

Lycos, 27

Maasai, 99

Macau, 22

Macedonia, 117

magnetic resonance imaging (MRI), 44, 52–53

Maguire, Eleanor, 44–45

majoritarianism, 168

Malaysia, 22, 77, 105, 132, 139

Mandarin, 124, 178

Mankosi, 100

Māori, 63–64

Marcus, Aaron, 105–6

marginalization: artificial intelligence (AI) and, 167–71; Bangladesh and, 161–68, 171–74; bias and, 168–77; Brazil and, 169; ChatGPT and, 168, 171, 200; China and, 165–66; colonialism and, 164–65, 170–71, 174–76; context and, 169, 172, 176–77; culture shock and, 5–6; developers and, 166–67, 173, 176; diversity and, 168, 170, 173; efficiency and, 164; Facebook and, 162, 166–69, 176; ideals and, 170, 176; identity and, 172–76; imperialism and, 13–14, 165, 167, 169; language and, 162–63, 169–77; localization and, 167, 169; majoritarianism and, 168; mobile payments and, 178–79; moderators and, 172–77; politics and, 162–65, 170, 175; religion and, 165–66, 169, 171–76; social media and, 166; Switzerland and, 175; Uber and, 161–67, 169; United Kingdom and, 171, 174; United States and, 163, 165, 169, 177; user interfaces and, 162, 169; values and, 13–14, 164, 167, 175; websites and, 173

Markus, Hazel Rose, 68–69, 71

Masakhane, 197

masculinity, 105–7

Massive Open Online Courses (MOOCs): Axim Collaborative and, 84; choice and, 78–86, 201–2; Coursera and, 78, 84; decision-making and, 78–86, 201–2; designers and, 78; digital utopianism and, 84–85; Edraak and, 78; education and, 78–86, 201–2; edX and, 78–79, 84–85; hype of, 78; iCourse and, 84; IITBombayX and, 84–85; Kentaro and, 85–86; motivations for accessing, 80; One Laptop Per Child Project and, 86; student-teacher ratios and, 81–83; Swayam and, 78; Udacity and, 78, 84; XuetangX and, 84

Masuda, Takahiko, 46, 50

McKinsey report, 75

McPartland, Thomas, 98–99

Mennecke, Brian, 95

Mercari, 21

Metaza-Kakavouli, Danae, 118

Mexico, 41, 57–58, 63, 114

Microsoft: Abokhodair and, 96; Clippy and, 103; context and, 40–42; Cortana and, 40–42; infrastructure and, 183; online communities and, 96; Teams, 183; Toyama and, 85; Windows and, 103

Midjourney, 167

Mixi, 30–32

Miyamato, Y., 50–51

Mkabela, Q., 189

mobile payment, 178–79

moderators, 172–77

Molinsky, Andy, 142, 194

MoneyFellows, 36

Mopan, 58

moral issues: ancient civilizations, 49; artificial intelligence (AI), 9, 154; BlenderBot, 154; choice, 9, 67, 194; equality, 14; freedom of choice, 67–68; infrastructure, 185, 192, 194, 196; robots, 9; social justice, 14, 182–87, 191; universal traits, 8; websites, 9

Moral Machine, 9–10

Morshed, Mehrab Bin, 162

Motsaathebe, Gilbert, 181

MSN, 124

mundane technologies, 180

Munoriyarwa, Allen, 181

Museum of Modern Art, New York, 121, 125–29

Muslims, 36, 96–98, 148, 166, 172–74

Myspace, 93

Mzungu, 61

Na, Jinkyung, 94

Naef, Myke, 62–63

Nairobi, 99

Namibia: Facebook and, 94–96; Herero and, 189–90; language and, 57–58; online communities and, 94–96, 100; ridesharing and, 164

national culture, 19–20, 107, 128

natural language processing, 152

Naver, 24–25, 26, 29

Neftlix, 75

neocolonialism, 13–14, 100, 165

Nepal, 57

Netherlands, 31, 39–40, 91

network effect, 30, 117

neuroscience: brain plasticity and, 12; cognition and, 44–45 (see also cognition); context and, 45; cultural, 51; information and, 44–45, 51; Kitayama, 51–52, 71–72, 74; Maguire study and, 44–45; research methodology and, 11

New Zealand, 63–64, 66, 117

Nigeria, 32–33, 82–83

Nisbett, Richard, 46–50

Nkwo, Makuochi, 32

NLP research, 197

Noble, Safiya Umoja, 180

noncontact cultures, 135, 143, 203

Norenzayan, Ara, 8–9, 63

norms: affective grounding and, 128; anthropologists and, 121, 138; artificial intelligence (AI) and, 127; attention and, 137; behavioral, 11, 15–17, 29, 71, 122–23, 131–38, 141–42, 194, 201; bias and, 20–21, 127; Canada and, 122, 130–31; ChatGPT and, 143; China and, 122, 125, 142; choice and, 125–28, 131, 133, 141, 202; cognition and, 136, 143; collectivism and, 132; context and, 121–25, 129, 132–39, 142–43, 203; designers and, 6, 12–13, 34, 126, 189, 193; developers and, 6, 12–13, 29, 133,

143, 166, 193; diversity and, 132, 142–
43; efficiency and, 120–22, 129; equality
and, 122; ethics and, 128; formal, 41,
120–23, 131, 185, 202; gender, 136, 139;
gestures and, 122, 126, 134, 136, 143;
Google and, 127; homogenization and,
127, 142; Human-Computer Interaction
(HCI) and, 126, 130; identity and, 125,
128; India and, 124, 128, 135–36, 139;
ingroups and, 144; intergroup threat and,
135; Japan and, 120, 123–27, 130–32,
135–37; LabintheWild and, 121, 138–
39; language and, 121–22, 124, 127,
131–32, 135, 138; lifeboat task and, 128;
misassumptions on, 203–4; othering
and, 15, 188; politics and, 138–39; psychologists and, 124, 128, 130, 135, 138;
religion and, 139; safety and, 139–44;
Switzerland and, 122; United Kingdom
and, 136; United States and, 122–29,
133–38, 141–43, 203; values and, 6, 12–
19, 51, 95, 122, 138, 148–49, 159, 164,
194–95, 198–201, 201
Norway, 33
Nyandeni municipality, 100

Oberg, Kalervo, 146
object-centered frame of reference, 57–58
Olacabs, 21
Oliveira, Nigini, 89–90
Olson, Gary, 123
Olson, Judy, 123
One Laptop Per Child Project, 86
online communities: autonomy and, 99;
    Brazil and, 94; China and, 89–92; choice
    and, 94; cognition and, 100; collec-
    tivism and, 89–99; colonialism and,
    99–100; color and, 92; context and, 29–
    34, 93–97, 101; cultural background and,
    92–96; designers and, 91, 93; education
    and, 89; efficiency and, 89–90; Facebook
    and, 89, 93–95, 100; Germany and, 91–
    92; Google and, 88; Hong Kong and, 89;
    ideals and, 157; identity and, 95–101;
    India and, 89–91, 94; individualism and,
    91, 94, 97–101, 202; innovation and,
    93; interdependency and, 95, 203; Japan
    and, 94–95; LabintheWild and, 98; Mas-
    sive Open Online Courses (MOOCs)

and, 78–86, 201–2; Microsoft and, 96;
    network effect and, 30, 117; power dis-
    tance and, 95; psychologists and, 98–99,
    112; Q&A communities, 89–92, 174;
    religion and, 88, 96–98; rewards and,
    92, 101; self-presentation and, 96–101;
    social media and, 96–101; Solidarity
    Economy and, 93; South Korea and, 91;
    Switzerland and, 91; Twenty-Statements
    Test and, 98–99; United Kingdom and,
    89, 99; United States and, 89, 93–95, 99;
    values and, 90, 95, 98, 100; websites and,
    100. See also specific group
Open edX, 84–85
Orji, Rita, 32
Orkut, 89
Østerlund, Carsten, 172
othering, 15, 188
outgroups, 25, 70–71, 76
Oxford Internet Institute, 170
Oyibo, Kiemute, 33

Pakistan, 81–83, 139, 168
Pang, Rock Yuren, 119
Papua New Guinea, 58
participatory design, 14, 190–92
Pelto, Pertii J., 18, 138–39, 141
Pepsi, 42
Personality Design Team, 41
Peters, Anicia, 95
Plantin, Jean-Christophe, 180
plasticity, 12
PNAS, 58
politics: borders and, 15; China and,
    21–22; communication and, 138–39;
    context and, 27, 30; culture shock and,
    150–54; infrastructure and, 181, 188;
    marginalization and, 162–65, 170, 175;
    norms and, 138–39
Pormpuraawan, 55–57
Portugal, 77–78, 169
postcolonial computing, 14, 188
poverty, 79, 85–86, 141, 168
power distance: communication and, 122–
    25; cultural background and, 105, 107;
    culture shock and, 149; decision-making
    and, 77, 83; misassumptions on, 203;
    online communities and, 95
priming, 50

privacy, 13, 22, 31–32
privatization, 180
Procter & Gamble, 42
profit, 7, 84, 132, 192
Prototype Nation: China and the Contested Promise of Innovation (Lindtner), 22
proxemics, 134–35
psychiatrists, 151
Psychological Review, 71
Psychological Science, 53
psychologists: acculturation and, 147; affective grounding and, 128; Arnett, 8; Barrett, 45; Berry, 147; Bochner, 145–46; Catholic Church and, 9n1; communication and, 124, 128, 130, 135, 138; cultural cycles and, 17; culture shock and, 145–47, 152; decision-making and, 62–71, 81; Furnham, 145–46; Fussell, 124; Gelfand, 138; Gutchess, 51–52; Hedden, 53; Henrich, 8, 9n1; information and, 45–46, 49–53, 66–71; infrastructure and, 195; intergroup threat and, 135; Iyengar, 70; Kitayama, 51–52, 71–72, 74; Lepper, 70; Levine, 62; limited research samples of, 8; Lindgaard, 112; Ma, 99; Mann, 66; Markus, 71; Masuda, 46, 50; Nisbett, 46–50; norms and, 124, 128, 130, 135, 138; online communities and, 98–99, 112; priming and, 50; Schoeneman, 99; Twenty-Statements Test and, 98–99; values and, 17, 138, 152, 195; Ward, 145–46; Western, Educated, Industrialized, Rich, and Democratic (WEIRD) people and, 9–11, 200; Witkin, 81
Psychology of Culture Shock, The (Ward, Bochner, and Furnham), 145

Q&A communities: Quora, 89, 91, 171–76; Stack Exchange, 89–92, 132, 174; Zhihu, 89, 91–92
Qadri, Rida, 168
Qatar, 97–98
QR codes, 179
Quora, 91; Bengali, 172–76; India and, 89, 171–74; Stages, 176
Quran, 96

racism, 180, 195–96
Raghavan, Prabhakar, 28
Randall, Dave, 180
rating systems, 103, 125, 162–63
reading direction, 39, 55
reading speed, 48
Reddit, 173–74
rejection stage, 147
relational mobility, 94–95, 202
religion: artificial intelligence (AI) and, 168; Bangladesh and, 165, 165–66; Buddhism, 173; Catholic Church, 9n1, *156*; censorship of, 166; Christianity, 153–54, 173, 175; communication and, 139; context and, 36; culture shock and, 148, 152–55; Facebook and, 166; health and, 166; Hinduism, 172–74; individualism and, 9n1; marginalization and, 165–66, 169, 171–76; misassumptions on, 204; Muslims, 36, 96–98, 148, 166, 172–74; norms and, 139; online communities and, 88, 96–98; subcultural groups and, 15; values and, 15, 148, 152–55, 175, 204
rewards: context and, 33; culture shock and, 145; decision-making and, 69, 76; financial, 8; online communities and, 92, 101
ridesharing, 7, 35, 163–64
rituals, 41, 96, 166, 198
robots: artificial intelligence (AI) and, 2, 9, 151, 153, 158, 195; behavior and, 3; chatbots and, 32, 151–57, 204; code-switching, 137–44; communication with, 133–44, 203–4; culture shock and, 151, 153, 158; elder care and, 130; ethics and, 9–10; high-context cultures and, 134, 142–43, 203; infrastructure and, 188–89, 195; low-context cultures and, 133–34, 137, 143, 203; moral issues and, 9; proxemics and, 134–35; robotaxis, 1–10, 35
Roesner, Daniela, 195
Romania, 116
Rouncefield, Mark, 180
Russell, Dan, 25–28
Russia, 40, 58, 80–82, 114, 116, 155
Rwanda, 20, 61, 86, 102–3, *104*, 107

safety: culture shock and, 159; norms
    and, 139–44; robotaxis and, 1–3; water
    quality, 139, 172
Samburu, 99
Sandvig, Christian, 180
Sato, Yumiko, 170
Saudi Arabia, 22, 32, 37–39, 97, 100
Schwartz, Barry, 18, 68
search engines: context and, 24–30, 37,
    41; cultural background and, 117;
    decision-making and, 76; impact of,
    180; infrastructure and, 180; racism and,
    180
secularization, 165
secular-rational values, 154
self-driving cars: context and, 201; Moral
    Machine and, 9; need for, 2; robotaxis,
    1–10, 35; safety of, 1–3
self-esteem, 147, 158, 200
self-expression, 69, 154–55
self-inflation, 72–73
self-presentation, 96–101
Semaan, Bryan, 95, 149, 172
SenseTime, 21
sentience, 153
Serbia, 117
Shelby, Renee, 168
Sheng, Hong, 179
silence, 122–23, 143, 203
Silent Language, The (Hall), 62
Silicon Valley, 20–21
similarity-attraction theory, 150
Singapore, 22, 77, 124
SixDegrees.com, 93
Slack, 120–24, 131, 143, 148
Smarter Screen, The (Benartzi), 76–77
Snapchat, 94
Snapdeal, 21
Sobh, Ibrahim, 5
social Darwinism, 9n1
social justice, 14, 182–87, 191
social media: context and, 30–31; culture
    shock and, 148–49, 157; decision-
    making and, 76; ideal life on, 157;
    identity and, 18, 31, 76, 97, 99, 202–
    3; influence of, 21; infrastructure and,
    182; marginalization and, 166; online
    communities and, 96–101
social proof, 64

SoftBank, 27
Solidarity Economy, 93
Solomons, 57
South Africa, 100, 181
South Korea, 22; context and, 24–27,
    37; Coupang and, 21; Google and, 24;
    information and, 47–48; KAIST and,
    47; Naver and, 24–25, 26, 29; online
    communities and, 91
Space Quest, 70–71, 75
Spain, 39, 78, 80, 116, 155
spatial frames of reference: absolute,
    55–59; egocentric, 55, 59, 94, 99;
    object-centered, 57–58
Spotify, 21
StabilityAI, 167
Stable Diffusion, 167
Stack Exchange, 89–92, 132, 174
STEM courses, 118
stereotypes, 13, 16, 62, 122, 130
Stern School of Business, 16
Stillwell, David, 94
Structural Magnetic Resonance Imaging
    (MRI) scans, 44–45
Strum, Christian, 4
Sun, Huatong, 193
survival, 94, 154–55
Swayam, 78
Sweden: context and, 42; cultural back-
    ground and, 116; culture shock and,
    155; decision-making and, 77, 84; Elec-
    trolux and, 42; KTH Royal Institute of
    Technology and, 193; Spotify and, 21;
    website preferences of, 116
Switzerland, 20; context and, 39; cultural
    background and, 102–8, 116; decision-
    making and, 62–63; JingDong and, 21;
    language and, 175; marginalization and,
    175; norms and, 122; online communities
    and, 91; website preferences of, 116

Taiwan, 21, 66, 68
Talhelm, Thomas, 73
Tamil, 124
Taobao, 75
taxis: infrastructure and, 179; "The Knowl-
    edge" and, 44; London cultural group
    of, 44–45; mobile payments and, 179;
    robotaxis, 1–10, 35; Uber, 161–67, 169

taxonomic terms, 27, 49, 59, 201
Tayama, Shinobu, 51–52, 71–72, 74
tech giants, 180
tech hubs, 20–21, 197
Thailand, 102–3, *104*, 107, 155
tight cultures: communication and, 138–42; culture shock and, 154; online communities and, 94
TikTok, 21
time: calendar software and, 88; clock, 62, 88–90, 202; decision-making and, 62–67; social, 62–63
Tiriyó, 58
Tocqueville, Alexis de, 18
Totonac, 58
Toyama, Kentaro, 85–86
Tractinsky, Noam, 109
TransferWise, 21
Triandis, H. C., 18, 66, 89
Trompenaar, F., 18
Twenty-Statements Test, 98–99
Twitter, 21, 96–97
Tzeltal, 57

Uber: Agarwal and, 161–62; Bangladesh and, 161–67; BRAC study on, 163–64; credit cards and, 161; cultural background and, 162–63; disruption from, 161–62; khep and, 164–65; language and, 162–63, 169; legal challenges of, 161–62; marginalization and, 161–67, 169
Udacity, 78, 84
Ukraine, 139, 155
uncertainty avoidance, 77–78, 106
Unicode Consortium, 127
United Arab Emirates, 22
United Kingdom: colonialism and, 99; context and, 31, 36; cultural background and, 117; culture shock and, 145, 155; decision-making and, 72–73; information and, 45; infrastructure and, 187–88; marginalization and, 171, 174; norms and, 136; online communities and, 89, 99
United States: colonialism and, 164; context and, 24, 27, 29–34, 37–41; cultural background and, 107–10, 117–18; culture shock and, 149–50, 153, 155; decision-making and, 62–84; hegemony

and, 200; individualism and, 19–20; information and, 46–53, 56–57; infrastructure and, 178, 181–83, 195–97; marginalization and, 163, 165, 169, 177; norms and, 122–29, 133–38, 141–43, 203; online communities and, 89, 93–95, 99; research methodology and, 11; robotaxis and, 1–10, 35; Silicon Valley, 20–21; Tocqueville on, 18; website preferences of, 117
Uruguay, 77
user interfaces: color and, 12, 25, 33, 76, 102–3, 105, 111, 184, 186; context and, 24–25, 27, 33; cultural background and, 102–11, 118; decision-making and, 76; information and, 54–55, 58–60; infrastructure and, 184, 186, 191, 193–95; marginalization and, 162, 169; minimalist, 24; perception of, 12; quantifying first impressions and, 111–17
US Social Science Research Council, 150
utopianism, 84–85
UX Pioneer, 105–6

values: anthropologists and, 15, 17–20, 138, 145; artificial intelligence (AI) and, 153, 195, 198, 200, 204; behavior and, 15, 17, 22, 71, 122, 153, 159, 194–95, 201; chatbots and, 153; ChatGPT and, 153, 195, 198, 200, 204; communication and, 122, 138; cultural, 6, 13, 18, 22, 106, 148–52, 159–60, 180, 194–96; cultural cycles and, 12, 17–20, 194; culture shock and, 13, 146–60; decision-making and, 81, 85; democracy and, 148, 152–53, 155, 204; designers and, 6, 12–13, 187–88; developers and, 6, 12–13, 187, 195–96; embedded, 198; equality and, 122, 148, 152, 155, 194, 204; ethics and, 196, 200; information and, 51; infrastructure and, 180, 187–88, 194–97; labels and, 16–19, 23, 168; majority's, 200; marginalization and, 13–14, 164, 167, 175; misassumptions on, 204; norms and, 6, 12–19, 51, 95, 122, 138, 148–49, 159, 164, 194–95, 198–201; online communities and, 90, 95, 98, 100; othering and, 15, 188; psychologists and, 17, 138, 152, 195; religion and, 15, 148,

152–55, 175, 204; secular-rational, 154; social, 138; traditional, 100, 154; World Values Survey and, 20, 152–55
"Value-Sensitive Design" approach, 188
Vashishta, Aditya, 149
Vassileva, Julita, 33
Venezuela, 92–93, 135
Venmo, 36
Vietnam, 63, 155
Vieweg, Sarah, 97–98

Wadi, Ahmed, 36
Wang, Lin, 134
Wang, Yuan, 125
Ward, Colleen, 145
Waymo, 1–2, 6
Waze, 21
Web Content Accessibility Guidelines (WCAG), 183–85
websites: bias in, 117–19; color and, 25, 33, 103, 105, 107, 111–19, 186–87; context and, 25, 27, 33–34, 37–42; cultural background and, 103–19; decision-making and, 78; designers and, 7, 34, 39–42, 54, 78, 100–19, 171, 183–87; information and, 47–50, 54; infrastructure and, 183–87; marginalization and, 173; moral issues and, 9; online communities and, 100; quantifying first impressions and, 111–17
WeChat, 21, 125–26, 178–79

Weibo, 89
Western, Educated, Industrialized, Rich, and Democratic (WEIRD) people: ethics and, 9; Google and, 29, 88; Human-Computer Interaction (HCI) and, 11; LabintheWild and, 23, 200; misassumptions on, 201, 204; Moral Machine and, 9–10; participant samples and, 9–11; problem of, 7–11; psychologists and, 9–11, 200
WhatsApp, 124
Wikipedia, 58, 117, 132–33, 169–71, 185
Wildlife Studio, 21
Winschiers-Theophilus, Heike, 95, 189
Witkin, Herman, 81
World Values Survey, 20, 152–55

XuetangX, 84

Yaaqoubi, Judith, 40–41, 182–83
Yahoo!, 26, 27, 29, 47, 89, 91, 124
Yamashita, Naomi, 48
Yeh, Tom, 111
Young, Iris Marion, 165–68
YouTube, 40, 75
Yu Garden, 178
Yugoslavia, 117

Zhang, Amy, 176–77
Zoom, 88, 183
Zuckerberg, Mark, 93–94